DO NOT REMOVE
CARDS FROM POCKET

COMMON SENSE TOXICS
IN THE WORKPLACE

COMMON SENSE TOXICS IN THE WORKPLACE

A Manual for Doctors, Nurses, Emergency Responders, Employers, Industrial Hygienists, Risk Managers, Claims Adjusters, and Lawyers

I. R. DANSE, M.D.

Associate Clinical Professor, Occupational and Environmental Medicine, University of California, San Francisco

President, ENVIROMED Health Services, Inc.
San Rafael, California

Consulting Physician in Toxicology, Pharmacology and Occupational Medicine

VNR VAN NOSTRAND REINHOLD
_____ New York

Copyright © 1991 by Van Nostrand Reinhold
Library of Congress Catalog Card Number 91–26646
ISBN 0-442-00154-1

Manufactured in the United States of America

Published by Van Nostrand Reinhold
115 Fifth Avenue
New York, New York 10003

Chapman and Hall
2-6 Boundary Row
London, SE1 8HN

Thomas Nelson Australia
102 Dodds Street
South Melbourne 3205
Victoria, Australia

Nelson Canada
1120 Birchmount Road
Scarborough, Ontario M1K 5G4, Canada

16 15 14 13 12 11 10 9 8 7 6 5 4 3 2 1

Library of Congress Cataloging-in-Publication Data

Danse, Ilene R.
    Common sense toxics in the workplace / Ilene R. Danse.
      p. cm.
    Includes index.
    ISBN 0-442-00154-1
    1. Industrial toxicology. 2. Industrial toxicology—Diagnosis.
I. Title.
RA1229.D36 1991   91-26646
615.9′02—dc20   CIP

To Kindness and Common Sense

To Jim

Contents

Preface

People have symptoms; people get sick. Which ones are related to the job, and how can you tell? *Common Sense Toxics* is based on my experience in a clinical toxicology practice. It suggests guidelines that simplify management of a person who is concerned about a "toxic exposure." It is not meant to replace textbooks of toxicology or medicine, and the reader should remember that each person concerned about "toxics" needs to be evaluated as an individual.

Just the very word "toxics" conjures up fear in the public—a public which includes physicians, nurses, emergency responders, employers, risk managers, insurers, lawyers, and workers. This book provides an approach to the treatment, differential diagnosis, and prevention of toxic exposures in workers. It simplifies the apparent complexity of claimed toxic injuries and explains why information obtained from laboratory experiments does not apply to workers. It provides the tools that a physician can use to decide whether a toxic reaction has occurred in a patient. It dispenses common sense advice for people who have to work with, or are responsible for, substances that have toxic properties.

Acknowledgments

The author gratefully acknowledges her patients and their experiences with toxic substances at work; excellent contributions by Dr. James C. Peterson, Linda G. Garb, M.D., Andrew C. Meuwissen, and Jacqueline K. Ross; time-consuming reviews, thoughtful comments, and editorial assistance of James Danse, Shawn King, Rosalie Moore, Arthur Raisfeld, Jean Meenaghan, Dr. Larry Strauss, John Cleveland, and the Honorable Jack Merrill; Julie Meehan for her assistance in preparation of the manuscript and Pam Touboul and the staff at Van Nostrand Reinhold.

Thank you, friends, for making this a better book.

COMMON SENSE TOXICS IN THE WORKPLACE

Part 1

A Short Course in Human Toxicology

1

Why We Need This Book?

KEY TOPICS

I. What is Toxicology?
II. How Can Common Sense Toxicology Help Persons Involved in Worker Health?

I. WHAT IS TOXICOLOGY?

Toxicology can be described as the study of harmful substances; this book is about substances that are harmful to people.

The trend in the field of toxicology has been to seek answers in laboratory rats and mice, but these studies miss the real point. They fail to illuminate human health effects and do not provide information that is helpful to today's worker.

II. HOW CAN COMMON SENSE TOXICOLOGY HELP PERSONS INVOLVED IN WORKER HEALTH?

In a busy toxicology practice, we frequently evaluate workers who are misdiagnosed. Medical records often come to us bearing a "diagnosis" of "toxic exposure" made by a general practitioner, psychologist, allergist, or "naturopath."

A 25-year-old woman has remained in a state of hysteria for four years, ever since she had a "toxic exposure" while picking fruit. She developed nausea and headache, which she attributed to pesticides from a crop sprayer flying in

3

the sky some distance away; the fact that she was working in a field adjoining one that was freshly fertilized with chicken manure was overlooked as a possible cause of nausea. She was hospitalized for observation. No evidence of pesticide intoxication was found, either by blood tests or by official investigators; however, she remains fixed in her beliefs and undoubtedly will remain so until her lawsuit against her employer is resolved.

A 22-year-old coworker was actually diagnosed as having pesticide poisoning when she had nausea and vomiting. It was only after several weeks and many doses of anti-anxiety, tranquilizer, headache and pain medication that her condition was accurately diagnosed: pregnancy, non-industrial.

Many patients receive medical advice to "avoid all chemicals." The story of what happens to these people prompted this book, because most of them lose their jobs. For example:

During an examination for life insurance, blood was found in the urine of a healthy 34-year-old pipefitter who had no symptoms. Several doctors advised him to "avoid all chemicals," even though they had no knowledge of prior urinalyses and ignored the fact that this man had essentially no exposure to chemicals in his work, which involved routine maintenance on the premises of a refinery. As a result of this advice, a highly-skilled worker lost a high-paying job. He also lost his trailer and boat, and borrowed money to meet his mortgage payments and provide for his wife and three children. As a result of his doctors' advice, he had no income for over a year.

A nurse, an editor, two computer programmers, a bookkeeper, a graphic artist, and five telephone operators have been "banned from indoor environments."

Surely, the toxics arena is very confused. Sometimes just the word "toxics" makes practitioners forget their medicine. They know that allergies in humans are specific. A person allergic to bananas should avoid bananas, not all fruit or all food! Similarly, a person sensitive to one chemical would not have to avoid all chemicals—each chemical is unique. A preclusion to "avoid all chemicals" can affect many lives, a whole person—a wage earner, a skilled worker, a valuable citizen—a person with a job. Another job may be very hard to come by. Physician-toxicologists, specialists in the field, are rare, and few physicians have had practical toxicology training or experience.

Where toxics are concerns, all must take seriously the responsibility to learn and respond appropriately—whether as physicians, nurses, emergency responders, risk managers, company health and safety personnel, industrial hygienists, claims adjusters, union leaders, attorneys or workers.

Basic concepts of problem solving require a statement of the problem, acquisition of facts, and analysis of them in order to reach a solution. Toxicology, perhaps more than any other field of medicine, calls for wisdom, discernment, and caution. It has not risen from the bowels of the earth with a special mystique steeped in witchcraft. It requires a little knowledge, some time, and everyday common sense. That is the theme of this book.

2

Everyday Toxicology

KEY TOPICS

I. OUR CHEMICAL BACKGROUND

Let us put ourselves in the place of the patient. What if we had a preclusion to "avoid all chemicals?" How would that affect a typical day? From the time we awaken, chemicals surround us—from toothpaste, shaving creams, deodorants, cleansers and powders, to coffee, fibers, and vitamins. All are "chemicals." Food is made of chemicals; organically grown vegetables synthesize high levels of their own chemical pesticides. Natural plant chemicals, when given to rodents, are carcinogens—in that they cause cancer—in the rodents. The distinction that chemicals intentionally added to foods are harmful, and that "naturally occurring" ones are safe, is illusory. Some of the most potent poisons, from hemlock to cancer-curing drugs, are made by plants (to protect them from predators) and ensure their survival.

At a lumber yard or home repair shop, all purchases—paneling, hardware, fabrics, etc.—are chemically treated. As a result, small amounts of chemicals are present in home air and in our bodies. They were measured by the United States Environmental Protection Agency (EPA) in an innovative study called "Total Exposure Assessment Methodology" or TEAM. In it, the EPA looked to see just what we are exposed to every day. TEAM studies identified and measured our chemical background in air, water, and breath. Ordinary people in blue-collar areas in New Jersey and California, residential areas in North Carolina, and extremely rural areas in North Dakota (population 25) volunteered to keep around-the-clock diaries of their activities and to wear monitors by which the air they breathed could be sampled. Daytime and nighttime personal air was sampled, as well as outdoor air and tap water. The outcome of this study was fascinating. The amounts of chemicals in outdoor air varied somewhat from one location to another, but, the quality of air in homes was pretty much the same, be it North Dakota or New Jersey (Table 2-1). All air contained some chemical "air contaminants." Air contaminants were present in minuscule or trace quantities—(measured in micrograms per cubic meter ($\mu g/m^3$)), amounts far too low to cause any effect on health.

People's exposure reflected their activities. For instance, perchloroethylene (PCE), a dry-cleaning solvent, was evident after visiting the dry-cleaner, while traces of benzene, in gasoline, were detected after filling up at a gas station. Benzene was found in both smoking and nonsmoking

TABLE 2-1   Some Chemicals Identified and Measured in Household Air in TEAM Studies, Average Values

Compound	Amount in Air ($\mu g/m^3$)
Benzene	9.46
Carbon tetrachloride	1.00
Chloroform	1.45
Dichlorobenzene	2.37
Styrene (vinyl benzene)	2.00
Perchloroethylene (PCE)	4.80
Trichloroethane (TCA)	14.00
Trichloroethylene (TCE)	1.15
Xylenes	19.46

Source: Wallace, Lance A. The Total Exposure Assessment Methodology (TEAM) Study, 1987

households but was higher in smokers' homes, since cigarette smoke is relatively high in benzene. Benzene is also naturally present in food and vegetation. Solvents other than PCE and benzene were detected after painting, or emanating from stain-resistant, or fire-proofed fabrics and furnishings, etc. It was no surprise that air contaminants were much higher in smokers' homes and breath than in nonsmokers. The only chemical that originated from water into air, was chloroform, derived from water disinfected by chlorine.

The point of all this is that every day, all the time, we all are exposed to trace quantities of chemicals. It is not possible to "avoid all chemicals"; nor are they all harmful. The amount that we are exposed to in everyday living, without an unusual incident, is known as "background." Besides measurement of chemicals in air and breath, background levels of some chemicals and their metabolic products, or metabolites, can be measured directly in people by analysis of blood, urine, or body fat. Background levels of many chemicals have been measured in trace quantities; none at levels known to cause harm.

Technology has become so sophisticated that extremely small quantities of substances, nearly molecules, can be detected. But, in spite of some attention-grabbing headlines, detection of chemicals in trace quantities does *not* mean that a situation is hazardous. Detection in the human body of trace chemicals does not mean that the person is "poisoned." Knowledge of typical background levels in the body, as discussed in Chapter 20, can help us determine if an "exposure" has occurred, as follows.

II. DEFINING AN EXPOSURE

An exposure occurs when a person receives a dose of a chemical at a level over background.

Lead can be used as an example of what is meant by an amount over background. Lead is a metallic element found naturally in the earth's crust and, for several millennia, it has been mined for numerous uses. Like the chemicals on the TEAM list, it does not occur "naturally" in the human body, yet all of us have some lead in our bodies from modern living — primarily from leaded gasoline and food — that, in small background amounts, is of no health consequence. The body is a pretty good toxic substances manager and can handle considerable quantities of potential toxins and drugs before harm occurs.

Lead is carried in the blood, largely in red blood cells (the erythrocytes). For lead, an idea of how much is stored in the body — the "body burden" — can be gleaned in several ways. The amount of lead in the blood can be measured directly (blood lead). An indication of lead burden can be gained by measuring its effects on blood proteins such as one called the "free erythrocyte protoporphyrin" (FEP). In cases of lead overexposure, the higher the FEP, the more lead there is likely to be in the body, and the longer it is likely to have been there. Lead may also be deposited and stored in bones or the nervous system, especially in cases of overexposure. Research techniques can be used to estimate the amounts of lead in these tissues.

What is considered to be the typical background blood lead level has decreased over time and continues to fall as the amount of lead in air, largely from leaded gasoline, decreases (Table 2-2).

Because people already have background levels of lead, what would constitute an excessive exposure? A lead exposure of medical concern results from direct contact with a soluble form of lead that gets into the body in excessive amounts through breathing or eating, is absorbed into the bloodstream, and shows up as a blood lead level significantly higher than current background levels. Whether that level causes harm is a *different* question.

Historically, when safety in lead mines was uncontrolled, lead miners frequently were poisoned. They felt tired, weak, developed headache (central nervous system dysfunction), severe abdominal cramps (lead colic), peripheral neuropathy, anemia and sexual problems at blood lead levels generally well above 100 μg/dL. Today, the most lead that a worker may have in his blood and still work with lead is prescribed and limited by law. The federal standards of the Occupational Safety and Health Administration (OSHA) (which are less stringent than those of some states) protect

TABLE 2-2   United States Public Health
Recommendations for Children

Year	Upper Bound Background Blood Lead Levels μg/dL
Mid-1960's and before	Less than 60[a]
1970	40[b]
1975	30[c]
1985	25[a]
1991	10 (anticipated)

a U.S. Department of Health, Education, and Welfare, Centers for Disease Control, 1985.
b Statement by Surgeon General of the United States for medical referral level, 1970.
c U.S. Department of Health, Education, and Welfare, Centers for Disease Control, 1975.

worker health and provide for items such as medical programs, blood tests, and safe work habits. "Medical surveillance" prevents lead poisoning. When blood tests indicate that a worker has a lead exposure, he is removed from working with lead before he is harmed. The work preclusion is specific—avoid lead.

III. BIOAVAILABILITY INFLUENCES EXPOSURE

Not all forms of lead pose the same potential danger. The presence of lead does not necessarily produce an exposure. Lead crystal, a decorative luxury in our society, must contain as least 24 percent lead in order to be called lead crystal. The higher the lead content, the more resonant the "ting." How can we drink plain water from a lead crystal glass? The lead in lead crystal is entrained in the glass and is relatively insoluble; it is not "bio-available" to us because it does not leach into the water in the glass. By remaining in the crystal, the lead is not able to be taken up by the human organism through the mouth, lungs, or skin; it is not bioavailable and cannot cause harm to the body solely by its presence. In contrast to water, other beverages such as acidic apple juice and alcohol may leach small amounts of lead into the liquid. Lead-glazed pottery is much more suscep-tible to leaching. It must not be used for foodstuffs because when it is in contact with acidic foods like orange juice or soda, significant amounts of lead *can* be released and dissolved into the food. Thus, the lead in this type of glaze would be bioavailable under certain circumstances and could lead to an exposure.

There are about six million other chemicals besides lead. People take a number of them—as drugs—as described in the next section.

IV. TAKING A DRUG IS A VOLUNTARY CHEMICAL EXPOSURE

If a person were given a work preclusion, "avoid all chemicals," does that mean that a person could no longer take his/her blood pressure medica-tion? Not even a baby aspirin for heart attack prevention? Of course not. Chemicals in the form of medication save lives and foster good health; the study of drugs is known as pharmacology.

Drug Dosages

Drugs are prescribed in doses measured in grams and milligrams (thou-sandths of a gram), taken several times a day, often for many days, some-times forever—as for high blood pressure. This intentional, pharmaceutical dose of a chemical is often thousands, even millions of times greater than trace quantities of chemicals found in home or work air.

Dose-Response Concept

Medication is prescribed according to a dose-response concept of both the benefit to the person and the risk of toxicity (toxic effect). In toxicology, the dose-response is also a critical concept. Trace quantities of most substances are not toxic, but at *some* level of exposure, there will be a reaction. Taking a quarter of an aspirin will have no effect on a headache, two tablets may relieve it, and taking a bottle constitutes aspiring poisoning, which may be fatal. As stated in the title of Alice Ottobani's wonderful book, *The Dose Makes the Poison.*

Drugs are chemicals intentionally used for a specific (beneficial) effect. The effects we do not want are the side-effects. The scientific study of drugs is called pharmacology, but pharmacology as a science cannot design a drug that is so specific that it has no side-effects. Drug side-effects *are* toxic effects. That is the bad news. However, because drug side-effects are toxic effects, doctors, pharmacologists, and toxicologists have acquired extensive knowledge about toxic effects in people. The good news is that what we have learned about the toxic effects of drugs in people can readily be applied to instances of chemical exposure in the workplace.

V. DRUG SIDE-EFFECTS ARE TOXIC EFFECTS

Depending on the drug, toxic or side-effects can occur in any organ or cell in the body. Take a drug that lowers high blood pressure for example. Besides reducing blood pressure, it may cause unwanted side-effects of dry mouth, drowsiness, depression, diarrhea, and impotence. These side-effects may be a high cost for a little lowering of blood pressure (these are the drug's toxic effects). When the drug is stopped, the side-effects disappear.

Physicians prescribe chemicals for people, and consumers (patients) take them. Indeed, the prescription drug industry alone is a multibillion-dollar-a-year business. In general, unwanted drug side-effects are common; but permanent effects from drugs are uncommon. Permanent effects from environmental or workplace toxic exposures are rare. The human body has a miraculous capacity to heal.

VI. PRINCIPLES OF THE ACTION OF DRUGS (PHARMACOKINETICS) AND CHEMICALS (TOXICOKINETICS)

Whether a person ingests a chemical in a pill or encounters a chemical at work, the way that the body handles them is the same. The fantastic voyage in a human (or animal) from entry to exit, is referred to as pharmacokinetics or toxicokinetics. These sciences are concerned with the following areas:

- Bioavailability: whether a drug can get into, and be active once it enters, the body
- Absorption: whether it can enter the bloodstream through the skin or gut, etc.
- Distribution: where it goes in the body
- Metabolism: whether the body changes it to something else
- Effect: the action that it has in the body
- Elimination: how the body gets rid of it

VII. DRUG ELIMINATION

A drug does not stay in the body forever. It reaches a concentration that is insufficient to cause an effect, and the effect ceases. Some quantity may still remain, but virtually all of it will be eliminated over time. Similarly, with "toxic" chemicals; their presence and toxic effects are usually temporary.

VIII. SAFETY AND CONTROLS OF DRUGS AND WORKPLACE CHEMICALS

Drugs in this country are regulated by the United States Food and Drug Administration (FDA) to assure the claimed benefits and to protect the public from undue risk of toxic effects. In the workplace, the Occupational Safety and Health Administration (OSHA, the regulatory agency), the National Institute of Occupational Safety and Health (NIOSH, the research arm of OSHA), and the American Conference of Governmental Industrial Hygienists (ACGIH, voluntary but respected standard bearers), have designated safe air levels of particular substances. These are known as Time Weighted Average Threshold Limit Values (TWA TLV), which are based on the concentration in air that could be inhaled over an eight-hour workday, five days a week, for an average working lifetime of 40 years.

The emphasis has been on clean air, but these groups need to devote more attention to skin absorption as a route of exposure. While some compounds are known to be absorbed through the skin or cause skin reactions, work practices that systematically avoid skin contact are not widespread, but should be, since fat-soluble solvents can be absorbed. For example, "glue or cement" for plastic pipe are solvents, which work by softening or "melting" the plastic. When area air samples and urine samples were taken from workers gluing plastic pipe, results showed that threshold limit values were not exceeded for air. However, because they used their fingers to apply the glue, solvents were absorbed through the skin and caused symptoms of intoxication in some workers (State of California Department of Housing and Community Development, 1989).

In summary, common sense tells us that the effects in the body of workplace substances, like drugs, are relative to the dose, depend on how the substance is metabolized or eliminated, and in reasonable quantities would not cause permanent injury or disability. In the next chapter, toxic reactions as measured in animal experiments and in workers are examined.

REFERENCES

Wallace, Lance A. 1987. The Total Assessment Methodology (TEAM) Study: Summary and Analysis: Vol. I. Office of Research and Development, U.S. Environmental Protection Agency, Washington, DC EPA/600/6-87002A.

Statement by the Surgeon General of the United States for medical referral level, 1970.

U.S. Department of Health, Education, and Welfare, Increased Lead Absorption and Lead Poisoning in Young Children. A Statement by the Center for Disease Control. 00.2629.

U.S. Department of Health and Human Services. 1985. Preventing Lead Poisoning in Young Children. A Statement by the Centers for Disease Control. 99-2230.

State of California, Department of Housing and Community Development, 1989. Draft Environmental Impact Report, Plastic Plumbing Pipe, Technical Appendices. Appendix C. Plastic Pipe Installation. Potential Health Hazards for Workers. Field Investigation FI-88-002, California Occupational Health Program.

3

Of Mice and Men: Toxic Responses in Laboratory Animals and Workers

KEY TOPICS

I. Animal Experiments
- Acute Studies
- Chronic Studies
- A Mouse is not a Man

II. Toxic Responses in People
- Duration
- Symptoms, Organs, and Chemicals
- Alcohol: An Example of a Toxic Response to a Solvent in a Person
  Dose-Response of Alcohol
  Acute Toxic Solvent Effect
  Other Acute Effects
  Reproductive Effects
  Cancer

Studies of toxic effects of chemicals used in the workplace can be done in two ways: by overdosing laboratory animals or by checking workers.

I. ANIMAL EXPERIMENTS

There are dozens of different species that can be tested, such as rats, mice, guinea pigs, hamsters, chickens, rabbits, dogs, fish, and so forth. Within each species there are dozens of subspecies and strains. Animals used for experiments are bred for this purpose, fed synthetic diets, and studied in laboratories under defined, artificial conditions.

Acute Studies

"Acute" experiments are quick, typically, a single overdose is given and death of the animal (or some other terrible response) is observed. Some experiments measure the lowest dose that kills an animal (LD_{LO}). Other experiments measure the amount of chemical that kills half the animals (LD_{50}). Unbelievably, the amount of chemical that kills half the animals, the LD_{50}, is used as an important parameter of toxicity. A chemical that kills the most animals at the lowest dose wins; it is considered the most "potent" or most toxic. However, a compound which is lethal to a guinea pig may have no effect in a hamster, so the circumstances under which the chemical is the most toxic are very specific. If a chemical kills a guinea pig and has no effect on a hamster, we still do not have information on the effects in people. In animal studies, besides species differences, toxicity varies according to factors such as age, dose, and strain. Toxicity may vary significantly according to route of exposure. What is meant by route of exposure is how the animal received it — by mouth, inhalation, injection, or rubbed on the skin.

Chronic Studies

From the acute experiments, the amount of chemical that does not kill the animal outright is determined so that longer-term studies may begin. A non-lethal amount, still an overdose, is given, generally for the life of the animal; rodents live for 1½ to 2 years. This type of experiment is known as a "chronic" study. Chronic studies are fairly straightforward: animals are dosed for their lives and studied to see exactly how sick they get. Animals that do not die are killed, and their organs are examined.

A special kind of chronic study is conducted to assess the effects of chemicals in rodent parents and several generations of their offspring. These are known as reproductive and teratogenic studies.

Most chronic studies are conducted to determine whether animals get cancer, but the way that these studies have been used accounts for much of the confusion in the toxics area. In many of the chronic or lifetime animal studies, benign or malignant tumors develop. As a result, there are long lists of "animal carcinogens." Few compounds, when administered to rats and mice bred for this purpose, do not cause cancers — in the rats or mice.

In comparison to animals, fewer chemicals are known to cause cancers in humans, but included among the known *human* carcinogens are our sex hormones, ingredients in birth control pills (oral contraceptives), and, ironically, treatments for cancer, such as chemotherapy and radiation. Cancer is such an important subject that Chapter 13 is devoted to it.

A Mouse is Not a Man

Animal experiments have obvious shortcomings in making analogies to today's work force. The dose and routes of exposure in animal experiments are vastly different from those of a workplace exposure. The toxicokinetics (Chapter 2) of the test animal may differ from toxicokinetics in humans. Furthermore, with respect to metabolism of chemicals, people really are individuals—one man may metabolize a compound like a hamster while another's resembles that of a guinea pig. Also, most animal experiments are conducted with overdoses of single ingredients while workers are exposed to small amounts of mixtures. Lastly, it must be emphasized that animals are overdosed because when amounts encountered in the workplace are given to them, it has no effect. When animal experiments simulate workplace situations, such as if a little chemical is splashed on a mouse or rat, or amounts of chemical usually found in workplace air are inhaled or ingested, there will be no observable effect of the chemical in the animal. Scientists, and especially toxicologists, do not like experiments where nothing happens. For years, it was impossible to publish experiments in which animals suffered no consequences from chemicals. This was a negative experiment, and no one was interested.

The hazards that excite animal researchers can be categorized as:

- Acute effects or damage
- Chronic effects
- Reproductive hazards
- Tumors or cancer

It is not possible to predict how a human will react just on the basis of high-dose animal experiments—people are not so easy to classify.

But what happens to a person?

II. TOXIC RESPONSES IN PEOPLE

Most symptoms associated with workplace exposures are acute, transient, and innocuous, but are not so readily classified as those in overdosed animals.

In people, toxic effects, somewhat awkwardly, can be categorized in several ways—according to how long they last (duration); according to symptoms (e.g., irritation); by the organ that is affected, for instance, skin;

or according to the class of chemical, such as "solvent." A brief overview is given here, and common complaints in workers are developed further in Chapter 7.

Duration

Onset of a symptom in a person may be sudden or gradual—a skin condition, like pimples, may bloom suddenly, or redness and scaling may develop gradually. Eye or respiratory irritation from chlorinated water in a swimming pool, from mace or household ammonia is a good example of an acute effect of irritants; once away from the pool, the mace or the ammonia, the eyes stop tearing.

In animals, a "chronic study" is long-term—for the life of the animal. In humans, "chronic" can have several meanings, so unless it is clarified, the definition is fuzzy. A chronic condition can be an effect, triggered by an acute event, that lasts a long time. (A wedding leads to the chronic condition of marriage.) In toxicology, a chronic response from a single incident is relatively unusual. Examples of chronic effects from an acute event are:

- A drug-induced lupus condition arising after a few doses of nonsteroidal anti-inflammatory drug for a painful shoulder
- Blindness from a splash of acid in the eyes
- Seizure disorder following lack of oxygen
- Aplastic anemia after exposure to the antibiotic chloramphenicol or the solvent benzene

In people, "chronic" may also refer to effects that continue because of ongoing (or chronic) exposures. Chronic effects from ongoing low-level exposures usually affect the skin and respiratory tract. Examples are epoxy resin-induced skin irritation (dermatitis) or eye irritation as a result of working with formaldehyde in close quarters.

Symptoms, Organs, and Chemicals

Irritants and dusts in the workplace commonly produce complaints of irritation of the mucous membranes of the eyes, upper respiratory passages (nose, throat, bronchial tubes), and irritation of the skin. Excessive inhalation or skin absorption of solvents may produce headache, dizziness, or nausea. Generally, symptoms affecting eyes, upper and lower respiratory passages, skin and the nervous system reverse when the exposure stops. These seem to be the most common symptoms in people, although in theory any organ can be affected.

Apart from eye irritation (measured by putting chemicals into animal's eyes) and specialized skin studies—mucous membrane irritation, headache and nausea are not symptoms that can be measured in animals. As can be seen in Table 3-1, some of the most common workplace complaints are watery or burning eyes, nausea, asthma, and hysterical or emotional reactions. Mice and rats do not develop rashes or complain of nausea and headaches; they only become hysterical when they see their next dose of chemical coming. For typical work-related complaints, information from people is required.

Alcohol: An Example of a Toxic Response to a Solvent in a Person

Workers often complain about solvents. To understand how solvents may affect workers, alcohol can be used as an example. In some workplaces, alcohol, as industrial ethanol, has replaced other solvents. However, vapors of industrial alcohols are irritating and drying, leading to mucous membrane and respiratory complaints. As both a typical solvent and a frequently used beverage, alcohol affords an excellent example of the entire spectrum of toxic effects—both acute and chronic toxic effects of solvents.

TABLE 3-1   Common Workplace Complaints by Category and Route of Exposure

| | Symptoms by Route of Exposure | |
Substance Category	Air	Skin
Irritating substances, e.g.	Watery or burning eyes	Redness
smoke	Drippy nose	Rash
soot	Burning skin	Burning
dusts	±Bad taste in mouth	
organic dusts	±Asthma (see index)	
ammonia		
Solvents, e.g.	Nausea	Defat
alcohols	Headache	Dry
glycol ethers	Dizziness	Crack
chlorinated	Drunkenness	Eruption
Reactive compounds, e.g.	Mucous membrane irritation	Dermatitis
epoxies	Pulmonary irritation	Rashes
amines	Sensitization	Hives
catalysts	Asthma	Blisters
isocyanates		
Odor awareness	Nothing	Nothing
	Nausea	
	Hysterical reaction	

±may or may not

Dose-response of alcohol. When confronted with a possible reaction to solvents at work, it is helpful to keep in mind the dose-response of alcohol when it is used as a beverage (Table 3-2). The relation between the dose of alcohol (in the left column) and the effect on the brain (in the right column) is presented in this table. In the case of alcohol, the dose-response is based on the fact that a particular amount in the blood corresponds to the known effects on the brain. This dose-response relationship provides the scientific basis for laws which define certain blood alcohol concentrations as intoxicating.

For alcohol, there are legal definitions of intoxication. In most states, a person is legally intoxicated at a blood alcohol concentration of 0.1 percent or less. This amount equals 1000 parts per million (ppm) alcohol in the blood, a quantity sufficient to impair driving ability in most individuals, as displayed in the column on the right (Table 3-2). As evident from the table, this blood concentration is reached, on the average, when a dose of 3 ounces or 30 grams of alcohol is ingested. Thirty grams, equivalent to an intoxicating dose, equals 30,000 milligrams or 30,000,000 micrograms. A small percentage of an absorbed dose of alcohol is eliminated in the breath. There is a relation between the amount of alcohol in the blood and the quantity in breath, providing a basis for breath tests that measure alcohol.

TABLE 3-2   Alcohol Dose, Blood Level, and Acute Central Nervous System Effects

Dose (Whiskey)	Blood Alcohol (Percent)	Acute Effects: Central Nervous System Depression
1 oz. (10 grams alcohol)	.015	Impaired visual acuity in abstainer "Rosy" world
	.035	Impaired driving Muscular incoordination
2 oz. (2 hours)	.05	Silly Labile emotions
3 oz.	.10[a]	Incoordination Increased reaction time for driving Changes on electroencephalogram (EEG)
	.20	Frankly drunk Slurred speech Double vision
	.30	Coma
	.40	Death (less with sedatives or other central nervous system depressants)

a. Legal limit for intoxication is between .08 and .10 in many areas.

Acute toxic solvent effect. Remember that even though some folks appear to be noisy and happy 'under the influence', alcohol *always* blunts or depresses brain functions. When centers of higher intelligence are suppressed by alcohol, more primitive brain centers take over. A person who gets drunk shows us an example of a really good, one-time, acute toxic reaction. Anyone who has ever had a hangover will tell you that they do not last forever—it just feels that way. At most, within a day or two, the effects of an acute overdose or toxic reaction to alcohol are gone. A hangover will not be present or worsen six months after a bachelor party or a convention.

Effects of solvents encountered in the workplace do not last forever, either; the same principles apply. A few drops of solvent on the skin or an odor of solvent in the air generally will not cause problems in a person. At work, inhaled doses of solvent are usually far less than that contained in the one or two beers that the worker may take after work is over.

Other acute effects. Besides the central nervous system, alcohol may affect other organs. Acute overdoses may produce gastritis (irritation of the stomach), altered blood fats, fatty liver, and abnormal heart rhythms. These toxic effects heal when alcohol exposure ceases.

Reproductive effects. Reproductive hazards, fortunately, are rare in U.S. workers, but alcohol has been associated with reproductive hazards in both men and women. In men, alcohol abuse leads to impotence and testicular atrophy, while alcohol abuse during pregnancy may cause the fetal alcohol syndrome which produces a damaged, retarded baby.

Cancer. One of the most feared of chronic toxic responses is cancer. Did you know that alcohol:

- Is a known human carcinogen?
- Is responsible for about three percent of human cancers?

Yet, many workers would gladly give up their work with solvents, but few their beer—even if presented with very convincing animal experiments or human data showing that the risk of developing cancer from beer was thousands of times greater than the risk of a problem from solvent use at work (Doll and Peto 1981; Ames, Magaw and Gold 1987). Cancer risks are further discussed in Chapter 13.

Now, it is time to examine some patients who have complaints of toxic injury.

REFERENCES

Doll, Richard and Richard Peto. 1981. *The Causes of Cancer.* New York: Oxford University Press.

Ames, B. N., Magaw, R., and L.S. Gold. 1987. Ranking Possible Carcinogenic Hazards. *Science* 236 (2) 271–280.

Part 2

How to Evaluate a Person for a Toxic Injury

INTRODUCTION

This section details an evaluation for a suspected toxic injury. To the usual medical exam are added descriptions of a job and toxic substance history, and a "toxic injury verification test." Common complaints, as observed in office practice, are discussed by symptoms and substances, and management of a person with a possible toxic injury is described.

4

Essentials of a Medical Examination

In a typical week, in a company town, seven workers from the plant appear with the following complaints:

- Marv complained of fever, malaise or being tired, the flu.
- Chicky had a rash on his forearms.
- Harold complained of shortness of breath.
- Tim was concerned about decreased sexual function.
- Bob, who had no complaints, had abnormal liver function tests on a routine exam.
- Roberto and Louie both complained of asthma.

How does the physician decide, within the constraints of a medical practice, who has a work-related injury?

Evaluation of a person for a toxic injury begins with a thorough medical exam. A comprehensive medical examination follows a two part convention: discussions with the patient (medical history) and "hands-on" portion (physical examination). Portions of the physical examination, such as measurements of height and weight; vital signs: temperature, blood pressure, pulse, respirations; or vision and hearing testing, may be recorded by a medical assistant. A checklist for the medical examination is followed by explanatory text.

I. Essentials of a Medical History
   A. Present ("Chief") Complaint (CC): Reason for the visit
   B. History of Present Illness (HPI): The (medical) story of why the person presented for examination. Should include:
      1. Present state of mind
      2. Medication history
         a. Prescribed drugs
         b. Other drugs
         c. Substance abuse
         d. Illicit drugs
      3. Allergic history/medication side-effects
         a. Medication allergies and side effects
         b. Other allergies
      4. Educational and social history
      5. Off-work activities
   C. Past Medical History (PMH):
      1. Childhood
         a. Illnesses
         b. Injuries
      2. Adult
         a. Illnesses

        b. Injuries
     3. Hospitalizations
     4. Surgeries
     5. Treatments or procedures
     6. Accidents
     7. Medications
     8. Allergies
     9. Physicians
         a. Past
         b. Present
  D. Family History (FH):
     1. Marriages
     2. Children
     3. Reproductive problems
     4. Diseases in immediate relatives
     5. Diseases in blood relatives
  E. Review of Systems (ROS):
     1. Nervous system
     2. Bleeding
     3. Memory
     4. Respiratory
     5. Infections/contagious
     6. Sexual
     7. Chest pain
     8. Heart
     9. Vascular
     10. Blood Pressure
     11. Abdomen/gastrointestinal
     12. Skeletal
     13. Muscular
     14. Hernias
     15. Genital
     16. Urinary
     17. Skin
     18. Mental

II. Essentials of a Physical Examination
  A. General Appearance
  B. Vital Signs (VS)
  C. Organ System Evaluation
     1. Head, eyes, ears, nose, throat (HEENT)
     2. Neck

   3. Lymph nodes
   4. Breasts
   5. Chest
   6. Lungs
   7. Heart
   8. Abdomen
   9. Genitalia
  10. Rectal
  11. Skin
  12. Skeletal
  13. Muscular
  14. Extremities
  15. Neurologic

I. ESSENTIALS OF A MEDICAL HISTORY

The medical history is the most important part of the examination. A good diagnostician will discern what is wrong with the patient about 80 percent of the time just by talking with him/her. This is more true than ever, even when sophisticated and expensive lab-tests are performed.

Obtaining a thorough history takes time. Filling out a medical questionnaire at home or in the doctor's waiting room can help to save both the doctor and the patient time by:

- Allowing the patient to write down what is bothering him/her
- Asking pertinent questions to jog his/her memory
- Allowing time to recall dates and events
- Indicating what information to bring, such as names of medication or workplace chemicals

A questionnaire also provides the physician with a permanent record prepared by the patient and provides the patient an opportunity to record his/her symptoms and complaints. In my consultation practice, patients fill out a 21-page medical and occupational history questionnaire. (Some, in answer to the question, "What makes you feel sad and depressed?" indicate, "Filling out this form," a response that is taken as a sign of good mental health.)

A history is essential for several other reasons. Patient and physician make their first contact by words — the astute practitioner senses that what the patient is complaining of is not necessarily what is bothering him. The doctor must uncover what is really behind the visit. For instance, a patient who is worried about having cancer might not come right out and say, "doctor, I am afraid that I have a cancer growing inside me," but might complain that he is losing weight because of his hay fever or allergy symptoms in the hope that the doctor will reassure him by agreeing with him.

Persons often visit doctors for reasons other than physical health — sadness, loneliness, life's stresses, and job dissatisfaction are examples.

In some specialties, a diagnosis can sometimes be made in moments — e.g., orthopedics, a fractured bone. In other specialties such as toxicology and internal medicine, time with the patient is needed to get a sense of the person and to pinpoint trickier diagnoses.

Present Complaint (CC)

The present complaint, sometimes also known as the chief complaint and abbreviated "CC," is the reason that the patient gives for visiting the doctor, for example, flu or asthma.

History of Present Illness (HPI)

The history of the present illness consists of circumstances leading up to the chief complaint. If the complaint is a ten-pound weight loss, the onset and duration would be established and information about diet, exercise, and appetite are obtained. It consists of a lucid series of facts telling a story leading up to this particular visit. If the visit was prompted by low back pain, did the back pain come on suddenly, during what activity, during lifting (appropriately or inappropriately), prior occurrences of back pain, general arthritic symptoms? Such clues allow the physician (who is really a medical detective), to obtain the factual basis for a correct diagnosis.

Present State of Mind

This is an extremely important part of the history. The present state of mind, or mental status, may appear in different portions of the written record. But, a person's state of mind is an important clue as to physical or mental distress and its cause. If a person is having stress in life — a vituperative divorce, a business failure, or the stresses of working, parenting, and householding alone — physical symptoms may result. Unless problems are recognized at the source, physical symptoms will not improve.

Medication History

Present medication. It is important to know what medications the patient is taking because what he/she is complaining of could be a medication side-effect. The more medication, the greater the chance of a reaction.

Drug interactions result from the prescribing of irrational combinations of drugs. These may consist of one drug antagonizing the effect of another, such as prescribing a drug to lower blood pressure (anti-hypertensive drug) with a nasal decongestant drug, such as pseudoephedrine or phenylpropanolamine, which elevate blood pressure, or other harmful reactions.

In general, the effects of drugs that act on the central nervous system are additive. The effects of alcohol, and alcohol in combination with other central nervous system depressant drugs such as pain medication, are also additive. The effects usually are temporary, but if enough drugs are taken over a period of time, alone or in combination, the effects can be permanent—there are some "fried brains" and "space cadets" around from the sixties. Patients on too many tranquilizers, and particularly the elderly, may develop similar symptoms.

The idea is that a very complete medication history is needed. Health care providers must be aware of all medicine a person is using from all sources, including alcohol and illicit drugs—not only because it is part of the drug history but because of the possibility of fatal drug interactions. For instance, a prescription for tricyclic anti-depressant medication to a cocaine abuser can be lethal.

Past medication history. Past medication history is also quite important. Persons who have been addicted in the past tend to get addicted again, and physicians must be on the alert not to prescribe potentially habit-forming or addicting drugs to them. This type of questioning requires rapport between physician and patient, which takes time to establish.

Allergic History and Medication Side-Effects

The allergic history includes medication allergies, drug side-effects, and other allergies, such as to foods, pollen, and insects. An allergic reaction to penicillin produced hives or cephalosporins caused a drug fever. A medication side-effect to codeine is nausea and vomiting. All are important to note.

Educational and Social History

This is background information: where the person was born, where and under what circumstances they were raised, educated, and lived. Some practitioners prefer to obtain a drug, smoking and alcohol history under the social part (with tact). Here, one notes daily cigarette consumption; whether smoking occurs at work and at the work station; illicit substance use; alcohol consumption; caffeine intake; AIDS risk factors and, if pertinent to health, religious practices.

Off-Work Activities

How a person spends non-working time can be particularly illuminating. The stress of the single working parent who commutes, works full-time, cares for four young children, and is struggling to make ends meet will become quickly apparent. Also, the patient who is too ill to come to a desk job but is continuing black-belt karate activities, jitterbugging, or is an entrant in the Indianapolis 500 will provide information that illuminates physical limitations. Second or third jobs, or income-generating activities around the home such as farming, ranching, or arts-and-crafts, are also noteworthy.

Past Medical History (PMH)

The style and organization of this section of the history varies from one practitioner to another, but the information to be obtained consists of childhood and adulthood illnesses, accidents or injuries, surgeries, treatments or procedures. Also, by asking a series of questions, the patient may recall treatments, procedures, medications, or allergies not previously given. The names, addresses, and phone numbers of past and current physicians may also provide additional information.

Family History

In this section the person is asked how many marriages they have had, how many children they have, whether there have been any reproductive problems, and whether immediate relatives or any blood relatives have particular medical problems. Some conditions such as asthma, essential hypertension, kidney disorders and diabetes run in families, so it is important to have a clear picture of the person's family. In this era of multiple marriages, family histories get more complicated than ever, so be sure to inquire whether they are giving you the family history of their children or a spouse's children from another marriage, which may not have any relevance to your patient's problem.

Review of Systems

Basically this is a list of symptoms, signs and complaints that helps to jog the physician's as well as the patient's memory. It covers the major organ systems of the body, such as the nervous system, blood, and respiratory, inquires as to infectious diseases, checks on the cardiovascular system, gastrointestinal system, musculoskeletal system, and genitourinary system as well as the person's mental status. By the time this series of questions is completed, either verbally or on a written questionnaire, the patient's memory is usually jogged well enough so that the history is relatively complete.

I also like to ask the patient whether there is anything we forgot to discuss, in case there was a problem that did not clearly fit into a category or there was a reluctance on the part of the patient to discuss something at the beginning of the interview.

II. ESSENTIALS OF A PHYSICAL EXAMINATION

This often begins with a general description of the person's appearance and demeanor, as well as a comment upon how willingly they presented the history, and often includes a comment as to the examiner's impression of how reliable or truthful the information obtained in the history was.

Vital Signs

This customarily consists of height, weight, blood pressure, pulse, respiration, and body temperature. Depending upon the reason for the examination, a vision or hearing test may be included. In most medical offices, a physician's assistant obtains and records this information.

HEENT

Examination of the head, eyes, ears, nose, and throat consists of looking at the size and shape of the skull, observing whether or not there is any trauma. Examination of the eyes not only checks the function of this part of the nervous system, but it is one place in the body where the physician can actually visualize the small blood vessels by looking in with a light, and as a result, diagnose conditions such as diabetes, high blood pressure, and certain infections. Examination of the ears, nose and throat can determine whether or not a person has "allergic-looking" mucous membranes which can be swollen and boggy, an allergy to aspirin associated with nasal polyps, or big tonsils that are interfering with respiration, for example.

Neck

Examination of the neck includes the arteries and veins leading to and from the head, lymph nodes draining the mouth, throat and head, and certain areas of the chest, the thyroid gland and cervical spine.

Lymph Nodes

Not only are lymph nodes located in the neck, and internally in the body, but externally they are found in the axillae (armpits), above the clavicles, and in the groin. Lymph nodes may be abnormal in certain infections such as mononucleosis, for example, or tumors such as lymphoma and lung cancer, or even drug reactions.

Chest

The chest examination may include the breasts in both sexes, the external and skeletal appearance of the chest, and then the lungs and heart within. The general shape and appearance of the chest is noted. It is tapped (percussed) to determine whether there is the right quantity of air within; it is palpated (felt) to determine whether there are any lumps, tender areas or unusual pulsations; and finally, it is listened to (auscultated) so that the sounds of the bronchial tubes, lungs and various areas of the heart can be heard. There are normal sounds and, as a rule, characteristic findings in disease states. In asthma, for example, which is a disease where the bronchi and bronchioles (airways leading to the air sacs of the lungs) narrow, wheezing would normally be expected to be heard during an attack. In a person who claims to have asthma, but where wheezing has never been heard in the chest by any examiner, the diagnosis would be questionable. The heart that is failing may be galloping at a rapid rate, while the lungs fill with fluid in pulmonary edema cases; these can be detected on the physical examination.

Abdomen

Like the chest, the abdomen is inspected, listened to for bowel noises and noises in the blood vessels (bruits), percussed for organ size and fluid, and palpated to outline the size and shape of the liver, spleen, kidneys and masses, if any.

Genital and Rectal Exams

These are usually part of a comprehensive physical check-up. The appropriate organs are examined. Doctors look for signs of infections, blockage, or tumors.

Skin

Skin is evaluated for color, temperature, appearance, texture, scars and lesions such as rashes, if any, which are then described in detail. Characteristics include location on the body part (e.g., only areas exposed to light), color (red, brown, purple), papular (raised) or bullous (large blisters), etc.

Skeletal and Muscular

While technically the skeletal and muscular systems are two separate organ systems in the body, they function together and are generally examined together. The shape and any obvious deformities of the skeletal system are noted. The appearance, feel and function of small and large joints of the body are tested by having the patient walk and/or move the appropriate parts. The extremities are examined for strength, tone, function, and circulation.

Nervous System

This includes various functions of the brain, cranial nerves I through XII and motor and sensory function in the peripheral nervous system.

Mental status, including orientation and memory may also be considered here. It includes comment about affect (whether they appeared depressed, sad, out of touch with reality, flat, cheerful, or distraught); orientation (whether they are aware of where they are); have normal memory; and whether psychiatric problems seem to be present.

The physical examination should be as comprehensive as possible. The patient should disrobe, and all major organ systems, as appropriate, should be examined. For example, the patient may be complaining of a work-related rash on the hands.

More than one case of a vesicular (blister-like) dermatitis on the hands has been erroneously diagnosed because no one asked the patients to remove their shoes, where severe tinea pedis (athlete's foot) existed. These were "-id" (me, too) reactions of the hands secondary to the condition on the feet. Carried to the extreme, we have observed a recent case of Von

Recklinghausen's Disease (neurofibromatosis). The patient had a hematoma on the chest, somehow initially attributed to pesticide poisoning. When his shirt was removed, it became evident that he was covered with fibrous tumors, some of which had reached golf ball size. One of these had ruptured, resulting in his problem. The patient recalled few physicians ever asking him to remove his shirt.

Not performing a thorough history or conducting a comprehensive physical examination can be embarrassing.

A patient being evaluated for an industrial injury must provide a very thorough job history which is discussed in Chapter 5.

5

Essentials of a Job History

A detailed job history is not something most physicians ordinarily bother with. In most of the forms that I see, there is often a short space after "occupation" in which a one-word description such as secretary, welder, baker, laborer, contractor, etc., can be filled in. There is little discussion of how and where the job is conducted. It is crucial to understand the details of a person's job in order to assess whether or not an exposure occurred sufficient to produce a toxic injury.

The essentials of a work history are as follows:

1. Employment history
2. Routine chemical contact history
3. Estimating opportunity for routine exposures
4. Industrial hygiene surveillance history
5. Occupational medicine surveillance history
6. Extraordinary chemical contact (incident history)

One very helpful way of obtaining this kind of information is to have the patient provide a narrative of a typical workday—"a day in the life of Bob." This type of narrative can be interrupted with questions. The work history will help to distinguish "contact" from "exposure." We have *contact* with substances we work with or near. An *exposure* occurs when that substance enters the body at a level over background.

I. EMPLOYMENT HISTORY

The employment history obtains the particulars about present as well as past jobs, such as date work began, lay-offs, lost time from work injuries, and, if appropriate, dates of termination.

- What are the person's overall job duties?
- How does he feel about the job, the employer, and the workplace?
- What is the workplace like?
- What is the type of industry?
- What are the known health hazards?
- Is the employer concerned about protection and prevention?
- What on-the-job injuries have there been?
- Has this worker had a pre-employment physical exam or interval medical surveillance?
- Have any work limitations or restrictions been imposed?
- If so, for what reason?

If the person is no longer employed, the reason for leaving that employer should be given. For example:

- Was it a disability retirement, an ordinary retirement, quit, fired, or other, and why?

- Was it for medical reasons?
- Was there a lay-off?
- Was it for a better job?

This type of information should be obtained for all jobs.

II. ROUTINE CHEMICAL CONTACT HISTORY

To learn what substances the worker has potential contact with, we need to learn about his job, and what he uses to do his job. The following series of questions may provide this information.

    A. What Does the Worker Do at His Job?
- What are the worker's job duties?
- What are the finished products of his job?
- What are the locations where the employee works?
- What tasks are performed at each location?
- How long does he stay at each task?
- Has there been a recent change?

For each task at each location:

    B. What Substances Does He Use?
- What are the raw materials?
- Have Material Safety Data Sheets been provided?
- Are there mixtures?
- Does he handle mixtures?
- Are there Material Safety Data Sheets for the mixtures?
- What are the process intermediates and/or streams?
- Are there by-products?
- What is off-gassed?
- Are there wastes?
- Are there mistakes?

    C. How Are The Substances Used?
- How frequently?
- How much?
- In what way are they used?

    D. How Is The Process Conducted?
- Is it an enclosed system with protective coverings?
- Is it done on an open workbench?

E. How Is The Area Ventilated?

F. Where Does He Do His Job?
- What shift does he work on?
- Who is his supervisor?
- During his shift, where is his supervisor located?
- How many other workers are in his work area?

G. How Does The Process That He Works on Function?
- What are the specifications?
- Does it usually function smoothly?
- Are there frequent malfunctions?
- Who is responsible for maintenance?

H. If a Particular Process is of Concern, When is This Process Done?
- Is it done every day?
- Is it done once in a while?
- For how long a time period is the process run?
- What is his role when this happens?

I. Has There Been a Recent Change?

III. ESTIMATING OPPORTUNITY FOR ROUTINE EXPOSURES

A. By Inhalation

We need to know what protective devices are in place, and whether or not they are used.
- If the worker is involved in a chemical process, is he segregated from the process?
    Does he work in an air-conditioned, computerized booth or cab?
    Does he stand over an open vat?
    Is the process completely enclosed?
    Are there fumes or fugitive emissions?
- What is the ventilation like in his work area or for each job task?
    Where is the intake of outside air located?
    Where is the exhaust?
    Are there purifying systems?
    Is there local exhaust or area ventilation?

Have there been mechanical failures?
Are there frequent breakdowns?

B. By Skin Absorption
 • In the workers' regular area, does he have skin contact in the normal course and scope of his activities?
 • What is his opportunity for skin contact?
   Does he routinely get material on his skin?
   Does he leave it on or wash it off?
   Are there unusual circumstances for skin contact?
   Does he shower before leaving work?
   What happens to the dirty clothes?
 • Does he use coveralls?
 • What does he use to prevent skin contact?

C. By Use of Protective Equipment
 • Overall, what protective equipment is assigned?
 • Of the protective equipment assigned, is it used?
 • If requirements for protective equipment exist, who sees that they are enforced?
 • When did the use of protective equipment commence?

Eyes:
 • What eye protective equipment is assigned?
 • What eye protective equipment is used?
 • Are contact lenses worn?
 • Are safety glasses worn?
 • Does he wear protective goggles?
 • Does he wear a face shield?
 • Other?

Feet:
 • Does he wear his own shoes?
 • What kind of footwear is it?
 • Does he wear leather boots?
 • Does he wear rubber boots?
 • Other?

Skin, Hands and Body:
 • Are impervious gloves worn for certain activities?
 • Does he wear his own clothes?
   If so, what?

- Is he given clothing by the company?
  If so, what?
  Does he wear a uniform?
  Does he wear coveralls?
  Does he wear a protective suit?

Respiratory:
  - What respiratory equipment is assigned?
  - What respiratory equipment is used?
    Does he wear a dust mask?
    Does he wear a cartridge-type respirator?
      If so, what type of cartridges?
      Who checks them?
      How frequently are they changed?
      Did he have a fit-test?
    Does he use supplied fresh air?
    What type of device?
    Does he work in an air-conditioned booth?

D. By Assessment of Work Habits
  - In his usual duties, is he neat?
  - Does he get covered with anything?
  - Does he eat or smoke at his work station?
  - Is there a potential for inadvertent ingestion?
  - Do any workers in his plant use drugs?
  - Have there been accidents or untoward occurrences not involving him?
  - Have workers left because of job-related illness?
  - Has he heard about any problems in the industry from coworkers or the union?

IV. INDUSTRIAL HYGIENE SURVEILLANCE HISTORY

Workers will often, but not always, be aware of an industrial hygiene program at their place of work. Additional information can be provided by the employer.

  - Are measurements of air contaminants conduced in his work area or elsewhere in the plant?
    If so, how often?

When was the last time?
Who has the information?
- Are there ovens, solvents or waste storage tanks?
- Are there fumes or odors?
  To where are the fumes or odors exhausted?
- Has there been area air monitoring at his job?
- Has he had any personal air monitoring in his breathing zone?
- Has he had any skin wipes?
- Has his work area had surface wipes?
- Has there been a special problem?

V. OCCUPATIONAL MEDICINE SURVEILLANCE HISTORY

Some workers are part of medical surveillance programs, that include examinations and tests provided by the employer. Some programs are mandated by law while others are voluntary.

- Is he part of an occupational medicine surveillance program?
  What are the details?
- Has he had any biologic monitoring (i.e., have levels of chemicals been tested for in his blood or urine, or on his skin)?
- Is his job or industry associated with health problems?

VI. EXTRAORDINARY CHEMICAL CONTACT (INCIDENT) HISTORY

Unusual incidents and accidents are important to record as overexposures may result.

- Was there a specific exposure?
- Did a spill occur?
- Did a release occur?
- Did something splash on his skin?
- Did a particular event happen out of the ordinary?

By the time this history is complete, the examiner should have a very good understanding of not only what Bob's job is, but how he does it, whether he is sloppy or neat, well-supervised, etc. This information will be used in assessing whether or not a toxic injury occurred.

6

Essentials of a Toxics History

Once you have a sense of the patient as a person and what his/her medical complaints and conditions are, and understand their job, it is time to take a toxics history. The essence of a toxics evaluation is provided by the answers to these questions.

1. What is the person's motivation for the visit?
2. What is the diagnosis?
3. Did the person sustain chemical exposure?
4. Was the exposure of sufficient intensity to produce an injury?
5. Is there evidence that an injury was produced?
6. Is the injury likely to be a result of the exposure?
7. Does the conclusion make sense?

We can call this the "Seven-Step Toxic Injury Verification Test."

43

I. WHAT IS THE PERSON'S MOTIVATION FOR THIS VISIT?

Besides the medical aspects, the motivation for a work-related claim must be understood. Unsavory motives for a claim of toxic injury capitalize on the general confusion in the toxics area and may be well disguised. Besides genuine concern over a toxic response, some other reasons for submitting claims are the following:

- Termination
- Employee/employer dispute
- Revenge
- Misdiagnosis
- Beating the system or fraud
- Plans for moving, career change, or retirement

Before reaching a conclusion, it is best to have a good sense of the main motivation for the visit, which may require time, skill, and insight to sort out. For example, some patients who are planning retirement wish to get a little disability allotment in lieu of or to supplement a pension.

> Take Daryl, a fellow who has worked for thirty years at the same job. He has had allergic rhinitis with heaviness in the chest for the last few years, which have been especially bad hay fever years. He has chronic bronchitis in the winter. Workers by law and by philosophy are now expected to know what they are working with, and he knows that, from time to time, he has worked with isocyanates. He believes or implies to his doctor that they are the cause of his chest condition. The doctor may not know much about isocyanates and if busy, he may not stop to question the facts. If Daryl and his doctor go back a long way together, facts may be "overlooked" to give Daryl and his wife a little "something extra" after all those years of hard work.

In my own medical training, Professors Lewis Thomas and Saul Farber, both of whom were Chiefs of Medicine during my education at Bellevue, repeatedly drummed into our heads, "Always listen to the patient." Indeed, it is a lesson I have never forgotten and will never forget. However, where toxics are concerned, I have to stretch this out a little: "listen to the patient, learn from him/her, get the facts, and reach your own conclusion." This can be accomplished by developing a rapport with the patient while taking a thorough history, followed by a comprehensive physical examination.

Some features of the medical history that suggest a physician should retain a healthy degree of skepticism when taking a toxics history are listed as follows:

- First physician visit after employment is over
- First physician visit following legal advice
- Employee/employer dispute
- Worker terminated
- Plant closed
- Impending layoffs
- Anticipated career change
- Anticipated return to school
- Already returned to (or accepted by) school
- Anticipated move to a new location
- Completed move to a new location

This list is by no means exhaustive. In the toxics area, we have found that patients can be boundlessly creative; one regrets that this energy and creativity were not used at their job.

II. WHAT IS THE DIAGNOSIS?

If a person complains of tearing eyes and runny nose and works with a potential mucous membrane irritant over an open dish, then the diagnosis is mucous membrane irritation. If a person has pneumonia, this is the diagnosis, and the cause needs to be determined. The differential diagnosis between medical and "chemical" symptoms will be discussed in more detail in following chapters.

III. DID THE PERSON SUSTAIN CHEMICAL EXPOSURE?

If the symptoms or findings are consistent with a possible toxic injury, this is the key question. The reason for this is that,

If there has been no exposure,
 there can be no injury.

Workers do not necessarily provide accurate histories on this point. They may not know what they are using, or may intentionally or unintentionally refer to it by the wrong name. Exaggerations are common, even when there is legitimate cause for concern. While, "Did the person sustain exposure?" boils down to whether or not a person received a dose of a substance over background (see Chapter 2), it will be arrived at by obtaining the chemical contact, exposure and incident histories as described in the previous chapter.

Estimating Exposure by History

Exposures can be estimated by determining the likelihood that a person received a dose of material over background during an incident — acute — or with regularity in the normal course of work duties — chronic, by the following routes:

- by inhalation,
- skin contact,
- improper use or lack of protective measures, or
- ingestion

Estimating Exposure from Tests

Tests that help to determine whether an exposure has occurred in the person (at a level over background) by examining samples from that person are known as biologic monitoring. Depending on the circumstance or substance, it may be possible to decide whether an exposure occurred after measurements in blood or urine samples. Specimens obtained pre- and post-incident can be analyzed to determine whether an amount over background has gotten into the body. More often than not, it is of immeasurable help in reassuring a person that an exposure did *not* occur. "Judy, even though some liquid splashed on your boot, none was detected on your skin or in your blood or urine." Chapter 20 is devoted to this subject.

If an untoward event has occurred, such as a spill, splash, or release, or if there is a strong suggestion of a work-related illness, a sample of blood, plasma and/or serum and urine *should* be obtained. If the content of the release is unknown, preserve specimens by freezing until it is known how they should be analyzed.

Examples of the types of tests that can help to illuminate a possible toxic problem are listed as follows:

1. Measurement from blood
   - Solvents/organic volatiles
   - Chlorinated pesticides or metabolites
   - Polychlorinated biphenyls
   - Metals, e.g., lead
   - Biological effects, e.g., cholinesterase inhibition
   - Therapeutic drug levels
   - Substances of abuse
   - Some metabolites

2. Measurement from urine
   - Metabolites of solvents
   - Metabolites of pesticides
   - Metals, such as arsenic and cadmium
   - Indirect effects of metals, e.g., B-2 microglobulin (for cadmium)
   - Drug and drug metabolites
   - Nicotine metabolites, such as cotinine

3. Measurements from skin

In an acute situation, especially if there is a spill or skin contact, a skin wipe can be obtained as the person is being decontaminated. At the present time, skin wipes are not in routine use but probably should be used for assessing exposures.

Other data that can be extremely helpful in assessing whether or not an exposure occurred are industrial hygiene air monitoring or surface wipe data, and records of occupational surveillance programs mandated by law for many chemicals. These can be obtained through a company's insurance claims adjuster or medical department.

IV. WAS THE EXPOSURE OF SUFFICIENT INTENSITY TO PRODUCE AN INJURY?

If a person was drenched with a chemical when a valve malfunctioned, the answer is "yes." If a person once saw an unopened can with a poison label on it in another building half a block away from where he usually worked, the answer is "no." More often than not, the answer to this question is a matter of good judgment. Remember that rodents receive enormous overdoses of chemicals and yet live to develop organ effects. In general, humans appear to be more resistant and do not receive such intense exposures.

V. IS THERE EVIDENCE THAT AN INJURY WAS PRODUCED?

Even given that someone was acutely or chronically exposed, was there evidence that something happened? Did he/she have a symptom of mucous membrane irritation, a rash, or a respiratory complaint? Oftentimes I see persons who are annoyed that they worked with chemicals and are sent with a diagnosis of "toxic exposure." This is not a diagnosis in my opinion, unless there was some objective manifestation of an injury. Objective evidence of injury with solvents could be eye irritation, nausea, or drunkenness—impairment of the central nervous system. Isocyanates may produce wheezing, and epoxy glues may produce rashes.

VI. IS THE INJURY LIKELY TO BE A RESULT OF THE EXPOSURE?

Would the symptoms mentioned above—the toxic injuries—reasonably result from the exposures cited? Dandruff, hemorrhoids, and ingrown toenails would not be reasonable manifestations of exposure to solvents, isocyanates, or epoxies. Nor would AIDS or pregnancy.

VII. DOES THE CONCLUSION MAKE SENSE?

The conclusion must make sense in order to be credible. If a person has a small bottle of typewriter correction fluid on his/her desk that is usually sealed and is taking ten tranquilizers a day, it is not likely that their feeling of sleepiness and sedation emanates from the correction fluid. It would make much more sense that these symptoms result from an overdose of prescription medication. While in theory both a solvent and a medication could cause a feeling of sleepiness, the quantity, the circumstances, and the exposure dictate that only one answer makes sense as a cause of the person's symptoms—the medication, not the correction fluid.

A true toxic injury will fulfill these criteria:

- Sound motivation for the visit.
- Symptoms consistent with chemical exposure.
- An exposure occurred.
- During exposure, enough was absorbed to cause injury.
- An injury resulted.
- The type of injury logically results from exposure.
- The scenario makes sense.

If it does not make sense to you, then it probably does not make sense.

No one is right all the time, although most physicians should certainly try to be. Especially in the toxics area, it may not be possible to know everything or have every last fact at one's fingertips. Therefore, let common sense be your very best guide.

7

Common Symptoms Associated with Toxic Substances

KEY TOPICS

I. Acute Symptoms
   - Irritants/Mucous Membrane Symptoms
   - Solvents/Central Nervous System Effects
   - Reactive Chemicals

II. Chronic Effects
   - Skin
   - Respiratory
   - Cancer

III. Symptoms by Substance Groups
   - Smoke, Dusts, Irritants, Fibers
     Acute effects: eye, mucous membrane, and respiratory irritation
     Chronic irritant effects: skin or pulmonary irritation from soot or dusts
   - Solvents
     Excessive inhalation
     Excessive skin contact
     Typical chronic exposure to solvents
     Central nervous system and liver effects
     Persistent liver function test abnormalities
     Skin effects
     Safe use of solvents
   - Odors
   - Pesticides and Agriculture
   - Inappropriate Protective Clothing

    Skin reactions
    Heat stress
    Job incompatibility
- Catastrophic Events: Highly Reactive Materials, Carbon Monoxide Poisoning and Oxygen Lack

It is not enough that a person saw someone work with a chemical, smelled an odor, or handled something with an unpleasant aroma. A toxic injury, in the workers' compensation sense, should have resulted in a disability that can be defined.

This chapter spells out symptoms and substances frequently associated with toxic injury. Chemicals in the workplace may produce acute symptoms coming on from single exposures, or those which result from chronic, low-level exposures over a period of time.

I. ACUTE SYMPTOMS

Acute symptoms are most commonly produced by irritants or solvents, or during a release.

Irritants/Mucous Membrane Symptoms

Irritant symptoms depend on the dose and the nature of the chemical; generally, there are no permanent after effects. Classic symptoms of mucous membrane irritation are red, watery eyes, runny nose and scratchy throat and lungs, for example, from a heavily chlorinated swimming pool. Symptoms last until the person leaves the pool, then disappear. Redness, smarting, and burning are common complaints of skin irritation. In addition to mere irritation, solvents may dry and defat the skin, predisposing to cracking, bleeding and infection (Table 7-1).

Solvents/Central Nervous System Effects

As a class, solvents affect the central nervous system the way alcohol does, and the dose-response is similar (Table 7-1 and Table 3-2). For alcohol, a teaspoon in a sauce has no effect, more produces a "rosy feeling," and a bottle consumed at once may be fatal. At very low levels of exposure, most solvents will not cause symptoms. But, if used sloppily, with poor ventilation or in enclosed spaces, conditions of excessive exposure, a spectrum of effects reflecting increasing central nervous system toxicity may result, ranging from a high, light-headedness, dizziness to drunkenness, to death.

TABLE 7-1   Common Workplace Complaints by Category and Route of Exposure

Substance Category	Symptoms by Route of Exposure	
	Air	Skin
Irritating substances, e.g.		
smoke	Watery or burning eyes	Redness
soot	Drippy nose	Rash
dusts	Burning skin	Burning
organic dusts	±Bad taste in mouth	
ammonia	±Asthma	
Solvents, e.g.	Nausea	Defat
alcohols	Headache	Dry
glycol ethers	Dizziness	Crack
chlorinated	Drunkenness	Eruption
Reactive compounds, e.g.	Mucous membrane	Dermatitis
epoxies	irritation	Rashes
amines	Pulmonary irritation	Hives
catalysts	Sensitization	Blisters
isocyanates	Asthma	
Odor awareness	Nothing	Nothing
	Nausea	
	Hysterical reaction	

± may or may not

If a man enters an enclosed space containing a high solvent concentration, without appropriate respiratory protection or enough oxygenated air to breathe, he may become drunk, unconscious, or die.

Take a slightly less extreme situation. A lot of solvent produces drunkenness, nausea, dizziness, or double vision. It is rare to have that kind of reaction in the workplace, so if someone is claiming loss of intellect from a solvent and he never had sufficient exposure to be "drunk," it is not likely to be a valid claim. As with a hangover, when the worker is removed from the source of exposure, these symptoms generally disappear, and there is no permanent injury.

In contrast to these dramatic and infrequent examples, there are occasional complaints of light-headedness or dizziness. If a person accidentally inhales some solvents, gets dizzy, and goes outside, that person should be fine by day's end or, at most, the next day. Accidental solvent inhalation should not result in six months or one year of lost time from work, yet such claims are common. Beware of claims of protracted symptoms from a relatively minor event; in these cases something is wrong with the allegation.

Reactive Chemicals

Catastrophic, highly toxic single releases, such as methyl isocyanate in Bhopal, India have been widely publicized. Reactive and highly toxic materials can produce chronic symptoms and lasting illnesses. In Bhopal, large populations caught in a toxic cloud, developed eye damage, pneumonitis, and pulmonary edema. If the potential for a toxic release exists in your neighborhood, the community, health care providers and the company must work as a team to ensure preventative safeguards and effective emergency procedures (Chapters 14 and 15).

II. CHRONIC EFFECTS

Continuing low-level exposures may produce chronic effects. Most commonly, the skin or respiratory system are involved.

Skin

Chronic skin complaints are common, often leading to disability retirements. Reactive materials such as epoxies, catalysts, or amines may irritate and/or sensitize. In contrast to dermatitis from pure solvents, where permanent disability is infrequent once the solvent is removed, permanent disability from chronic skin reactions *is* common. To avoid chronic and disabling skin conditions, more attention must be paid to good work practices that avoid skin exposure. Reactions may be prevented 1) by procedures that eliminate or minimize skin contact, such as using a tool instead of a finger, 2) by prompting washing after skin contact and 3) by early recognition and intervention.

Respiratory

Some substances sensitize the airways and lungs, and thus produce industrial asthma. The typical industrial asthma patient wants the job but develops tightness in the chest or trouble breathing whenever the chemical offender (e.g., isocyanates), is inhaled. Isocyanates are so predictably a cause of asthma in some workers that these chemicals must be used in enclosed systems or with air-purifying respirators. An individual with isocyanate-induced asthma, once sensitized, cannot return to any job where isocyanates are used (e.g. auto body spray-painting) and generally is entitled to vocational retraining.

Cancer

Cancer, an infrequent but dreaded chronic condition is uncommonly caused by work. However, it is a subject of such importance that Chapter 13 is entirely devoted to it.

III. SYMPTOMS BY SUBSTANCE GROUPS

Classifying substances by categories familiar to workers, along with common complaints, organizes this information for easier assessment. Common categories of substances encountered in the workplace are the following:

- Smoke, dusts, irritants, fibers
- Solvents
- Odors
- Pesticides and agricultural products
- Inappropriate protective clothing
- Highly reactive materials
- Oxygen lack

The category, conditions of exposure and acute and chronic symptoms for these substances are listed in Table 7-2.

Smoke, Dusts, Irritants, and Fibers

Acute effects: eye, mucous membrane, and respiratory irritation. All of us have had watery eyes from an onion, but no one views this as anything more than a transient irritation. The workplace equivalent of this kind of symptom is listed in Table 7-2. In a workplace, smoke, dusts, soots, and irritants could cause symptoms, such as smoke inhalation following a workplace fire or eye irritation from ammonia cleansers. With irritants, symptoms are usually transient and do not require lost work time, although it may be an annoyance to the person and require a doctor's visit. Generally, by the time the person visits the doctor, symptoms have disappeared, although sometimes irritation or redness of the affected mucous membranes remain. Generally, a patient who has had an irritant response just needs *reassurance.*

Chronic irritant effects: skin or pulmonary irritation from soot or dusts. Some dusts, such as those arising from repeated inhalation of soot, wood dust, or bean dust without respiratory protection can lead to chronic symptoms. Examples would be skin irritation of chimney sweeps or Wil-

TABLE 7-2   Presentation of Some Workplace Complaints by Category of Substance and Complaint

Substance Category	Conditions of Overexposure	Main Complaint	Symptoms and Findings in Relation to Exposure	
			Acute	Chronic
Smoke, dusts, irritants, fibers, cleansers and disinfectants	High dose low toxin, low dose reactive toxin	Irritation: Eye Mucous membrane Respiratory Skin	Brief, transient, temporary symptom. Compare to household ammonia, chlorinated swimming pool, barbecue smoke	Respiratory relatively uncommon Chronic bronchitis or sinusitis Skin relatively common. Chronic dermatitis.
Solvents	Work without respirator in confined space	Dizzy Nausea Headache	Systemic: Brief	Systemic: Infrequent unless major overexposure. Compare to alcohol.
	Excess skin absorption Sloppiness	Skin problems	Solvent: Defatting of skin, cracking, fissuring Gloves: May cause allergy or rash from sweating	Dermatitis Chronic skin complaints common when solvents used as glues or cements with plastic, rubber, epoxy, or in coatings containing them, and with direct contact. Type of reaction influenced by substance being dissolved

Odors Varied — universe of substances	None Nausea Hysteria "Feeling of being poisoned"	Some fuss, develop hysterical posturings. Compare to passing a feed lot in your car, smelling fresh paint, using spot remover.	Injury: none
Pesticides, organophosphates May be absorbed through skin or inhaled; rapid onset between exposure and symptoms (minutes to hours)	Cholinergic crisis	Symptoms are acute and transient. Increased secretions, salivation, sweating, respiratory distress, wheezing, pinpoint pupils. May be specific antidote. Complete recovery.	Uncommon. Consider another diagnosis.
Protective clothing (inappropriate) Too hot, too much clothing, sweating underneath	Emergency: Heat stroke, too hot	Heat stroke or heat stress disorders	None Treat for underlying condition
	Rashes	Localized or generalized bacterial infections may be present, contact dermatitis from rubber gloves Boots worn aggravate Athlete's foot	Provide suitable gear to prevent chronic skin reactions
	Panic, discomfort	Psychological intolerance Inability to tolerate clothing or respirators	No injury. Person or gear may be suboptimal. Re-evaluate.

TABLE 7-2 Presentation of Some Workplace Complaints by Category of Substance and Complaint (*Continued*)

Substance Category	Conditions of Overexposure	Main Complaint	Symptoms and Findings in Relation to Exposure	
			Acute	Chronic
Highly reactive materials	Release or inhalation, highly reactive materials, reactive intermediates, poison gases	May be extreme emergency Respiratory Eye Flu Fever Skin Sick	Acute illness, sick, severe Support vital signs; decontaminate Prevent anoxia, tissue necrosis if possible Airways and lung tissue, eyes can be damaged, i.e., methylisocyanate (Bhopal). Use specific antidotes when possible Assess lung with DLco	Acute injury may produce chronic symptoms, depending on conditions.
Carbon monoxide poisoning	Inappropriate combustion in indoor environment; improper engine or motor exhaust; fires	Emergency: carbon monoxide poisoning	Administer oxygen, support vital signs Specific treatment according to level of coma Symptoms vary from mild to severe.	Residual related to severity and circumstances of acute injury
Lack of oxygen	Low oxygen environment, with or without dusts, vapors, fumes, inert gases; respirator errors	Emergency: Oxygen lack	Acute anoxia (lack of oxygen) May be superimposed on noxious fumes or vapors Oxygenate rapidly	Residual would be due to anoxic damage and other substances present.

liam Bendix shoveling coal in "The Hairy Ape." In the lung, excessive exposures may cause permanent damage. The effects of cigarettes afford a good dose-response example of chronic irritation. It takes a lot of cigarette smoke every day for many years, inhaled directly into the lung of the smoker, to produce an objective change. Even so, when the exposure stops and the person stops smoking, the damage can reverse, up to a point. Lung damage from occupational exposures needs to be distinguished from that of smoking (Table 7-2).

Fibers of any kind, including asbestos, although inert, linger in the body and stimulate inflammatory reactions—chronic irritative responses that may eventuate in tumors (Chapter 13). In contrast to the lung, where exposures are generally asymptomatic, skin exposure to fibers such as fiberglass may cause itching. Secondary irritation may develop from scratching.

Solvents

Low levels of most substances such as common solvents do not produce any effect, while large amounts may be harmful (Table 7-1 and Table 7-2).

Excessive inhalation. Although many industrial solvents are not now intended to be taken internally by humans, several have been used as drugs. Such knowledge proves invaluable when evaluating toxics claims concerning them. For instance, trichloroethylene is an old inhalation anesthetic which is still being used today in some places.

As a category, solvents usually do not cause symptoms unless used inappropriately. This occurs when excessive amounts are inhaled in an enclosed space or when the wrong mask or respirator is used. Excessive exposure through the skin, respiratory system, or even ingestion (remember alcohol) may produce the following symptoms:

- Dizziness
- Nausea
- Headache
- Drunkenness

Excessive skin contact. Even if air levels are within the threshold limit values, absorption through the skin can be a source of undue exposure. The absorbed dose (the amount that a worker takes into the body) may vary appreciably from one worker to the next. A neat worker who avoids getting any solvent on the skin, or who washes it off promptly, will absorb little or no solvent in comparison to a sloppy worker with poor hygiene who wets their clothing and skin and remains that way for extended periods of time.

Typical chronic exposure to solvents. A candidate for excessive skin and inhalation absorption is a worker who is sloppy or hurrying and customarily does the same job day after day in a small, enclosed space and with excessive skin contact, getting it on the hands or skin. With the increased use of plastic pipes (such as for plumbing), which are glued together (the "glue or cement" being a solvent that dissolves plastic), laborers who glue pipe on a daily basis in confined spaces without fresh air or tools that help to avoid skin contact may be affected.

Central nervous system and liver effects. Transient liver function test abnormalities have been described after overexposure by inhalation or skin absorption sufficient to cause central nervous system (CNS) depressant effects of dizziness, nausea, headache, and/or drunkenness. Apart from overexposures, such as when gluing plastic pipe, it is our experience that liver injury or abnormalities in liver function tests occurring from incidental use of solvents (e.g., every few days to clean parts or from the occasional use of a vapor degreaser) do not occur.

Persistent liver function test abnormalities. Incidental solvent use at work, in our experience, like an occasional glass of wine or beer, does not result in liver function test abnormalities. Other, non-work-related, causes are more likely, such as prescribed drugs, alcohol use, obesity, lipid abnormalities, or liver disease. Unless a solvent has caused some symptoms, aberrations in liver function tests detected on routine screening should not be attributed to incidental solvent use. Even when excessive exposure produces a transient aberration in liver function tests, ordinarily this disappears as soon as the person is removed from exposure. Consider the effects of alcohol on the liver. A moderate amount of exposure, i.e., one or two beers a day, will not alter liver function. Consumed to excess over weeks, months, or even years, transient abnormalities in liver function may occur, but when the exposure stops, test abnormalities usually reverse.

Many Material Safety Data Sheets from solvents describe liver and kidney injury in intentionally overdosed mice and rats, but these symptoms do not occur in asymptomatic humans during ordinary use of solvents at work (or at home).

Skin effects. Systemic (internal) effects from solvents, as noted, are quite uncommon, but acute and chronic skin changes are frequent. Solvents *do dissolve* fatty substances — that is why we use them — and also dissolve the natural lubricating oils and fats of the skin. Two alcohols, isopropyl (rubbing alcohol) and ethanol ("alcohol"), are solvents. Both will dry the skin when applied repeatedly. While intact skin is an effective barrier, when skin is cracked and dry from solvents, the normal protection is disrupted and solvent can be absorbed. Injured skin is also susceptible to irritation and further damage. Most assembly line or highly skilled workers value their jobs, and accordingly, take very good care of their skin, because a solvent dermatitis interferes with the use of their hands, and thus, with their ability to work.

Safe use of solvents. Impermeable gloves of material resistant to the solvent being used may be appropriate for some activities. Impermeable gloves cannot be worn for protracted periods of time, because excess sweating under gloves will also produce skin irritation (dermatitis). Protective gloves are not inert—allergic reactions to glove fabric may develop. However, gloves will prevent skin contact when used appropriately, such as for cleaning tools or parts. They must not be used when fingers may get caught in moving machinery; a brush on a handle should be used or the machine disassembled. Gloves cannot have direct contact with glues—both get gummy—applicator tools must be employed.

Solvents, rarely used alone, are used to clean oil or dirt from parts or dissolve something else, such as plastics, rubber, paint, coatings, and so forth. The substances being dissolved may contain reactive materials such as constituents of epoxies and/or sensitizing materials like amines that are found in natural and synthetic rubbers. As several components may produce reactions, sometimes a mixture, not a single ingredient, is implicated as a cause of symptoms.

Two very simple guidelines prevent most chronic reactions. First, the skin should be protected by using clothing or tools provided by the employer, thereby avoiding skin contact. Secondly, employees need to be trained so that if any materials do get on their skin, prompt removal is followed by a thorough washing.

Odors

Several large groups of persons have been observed as well as a number of individuals who, as far as we could tell, smelled a bad odor at work one day and never came back. Some of these individuals had submitted claims of total disability and incapacity from life. According to investigations, the offending odors emanated from causes such as cooking oils, pesticides, paints and lacquers, carpeting, fertilizer, and perfume. In no case was there any evidence of any intoxication apart from an odor. Some smells, however, seem to trigger mass hysteria by nauseating or frightening people. In several large groups, about ten percent of the work force became hysterical while most other employees remained at work without incident.

Because odors sometimes seem to trigger mass hysteria, they should be dealt with by an overreaction of concern and reassurance. It is also wise, if there are certain odors that are unpleasant (such as fresh chicken manure as fertilizer), to wait until the odor dissipates before returning a work crew to the area.

Pesticides and Agriculture

There appears to be an inverse relationship between the degree of hysteria over pesticide odors, especially in indoor environments, and symptoms of intoxication, as can be seen in these case examples:

A 27-year-old pesticide applicator wet his clothes with an organophosphate cholinesterase-inhibiting pesticide while exterminating the perimeter of a large building. Though much of his clothing was drenched, he worked (contrary to proper procedure) for several more hours to complete the job, then showered at home. By then, he felt ill and was hospitalized with acute organophosphate intoxication. By the next morning, he felt well although a little shaken, and returned to work within a few days.

On a Monday morning an office worker was displeased by the smell at her work station. The preceding Friday a pesticide was applied for carpet fleas. Although she was relocated to several other work areas on a different floor, she subsequently claimed that rashes, dental trouble, bad teeth, and a substance abuse problem were due to this incident.

A 27-year-old fruit picker was nauseated by an unpleasant odor as she worked in the field. She noticed an aerial crop sprayer several miles away and assumed that what she was smelling must have been from the crop sprayer. She left the field and has not worked for three years, but through an attorney, has submitted a claim for incapacitating disability. No one seems to have made an association between the odor that she smelled and the chicken manure freshly applied to an adjoining field on the day that she complained of the odor. This was a chicken manure case.

Inappropriate Protective Clothing

Skin reactions. Skin reactions to protective gloves and boots is increasing as their use increases (see discussion on p. 59). Skin disorders may accrue from allergies to glove material, from increased sweating or secondary infections.

Heat stress. Impermeable gloves are not the only protective gear causing problems. With the *anschluss* of encapsulating suits and impermeable clothing, heat stress and skin disorders associated with sweating or infection are increasing.

A person forced to wear a rubber suit in a 110-degree sun developed heat stroke.
Workers in three layers of impermeable clothing on a "hazardous waste" site all developed folliculitis (bacterial pimples).

Job incompatibility. Protective gear can produce other problems. A job that appears straightforward in ordinary clothing may become dangerous in a moon suit with a fresh-air-supplied respirator.

A worker was rendered infertile as a result of exposure to dibromochloropropane (DBCP) during its manufacture. The company was a good company, and he and his coworkers chose to remain at their jobs. Then, nothing was spared in the way of protective equipment, and most workers wound up in encapsulating suits supplied with fresh air. Shortly thereafter, while obtaining a sample from the top of a tank car, he tripped over the air line, fell off the tank car, and fractured his back—a reminder that protective equipment is only protective if it is compatible with the job.

Inappropriate use of respirators can be fatal, as described in the next section.

Catastrophic Events: Highly Reactive Materials, Carbon Monoxide Poisoning and Lack of Oxygen

We have discussed common complaints that a physician may see in office practice in this chapter. Depending on severity, some may be encountered as emergencies. In this section we discuss acute emergencies likely to be dealt with on-site and at hospitals. These tend to fall into three groups: damage from highly reactive materials, carbon monoxide poisoning, and injury from circumstances resulting from a lack of oxygen. Highly reactive materials were discussed earlier (see Table 7-1). Specific antidotes and emergency life sustaining measures may be required (Table 7-2).

Carbon monoxide is still such a common form of poisoning that it is worth singling out for a moment. Fortunately, most cases of carbon monoxide intoxication do not occur in workplace settings but rather result from generation of incompletely burned materials as in fires, faulty use of equipment to heat homes such as improperly ventilated coal-burning stoves and kerosene heaters, and exposure to combustion engines and vehicle exhausts. Carbon monoxide poisoning occurs in workplace settings under the same circumstances. Exposure to increasingly high concentrations may produce headache, weakness, dizziness, nausea, confusion and unconsciousness. These effects are due to the ability of carbon monoxide to interfere with the oxygen-carrying locations on red blood cells, thus depriving vital tissues, such as the brain and heart, of a needed supply of oxygen. Carbon monoxide poisoning should be considered as a cause of symptoms in someone who is working in the vicinity of sources of combustion.

Unfortunately, deaths and serious injuries still result from circumstances promoting a lack of oxygen. These include unprotected entry into poorly ventilated, oxygen-poor, enclosed tanks, with or without fumes of an intoxicating substance. Unfortunately, this type of injury has occurred with protective equipment improperly used. Of occupational deaths or catastrophic illness, a number have been due to improperly connecting respirators to inert gases such as nitrogen or argon instead of to oxygen. Clearly, this type of situation is untenable.

8

Differential Diagnosis of Common Respiratory and Flu Symptoms—Medical or Chemical?

KEY TOPICS

    I. Cough
   II. Hay Fever, Allergic Rhinitis
  III. Chemical Irritation
  IV. Asthma
      • Non-Work-Related Asthma
      • Occupational Asthma

    V. Pulmonary Function Tests
  VI. Laryngospasm, Angioneurotic Edema
 VII. Flu Symptoms

In a busy January, more than half the patients in a physician's office may be there because of respiratory symptoms. It is a safe bet that most of them will not be there due to workplace factors. How does a busy physician who has many constraints on his/her time determine which one of them has a physical symptom related to an on-the-job exposure? First, the key word is TIME—it takes time. In order to do it right, a doctor has to stop, think, and sort out the facts. One reason is that, generally, symptoms from a "toxic exposure" are not specific. A body has just a limited number of ways of responding. For example, both a cold and a mucous membrane irritant will produce watery eyes and a runny nose. Let us see how common medical problems—upper respiratory infection (URI) symptoms, asthma, and flu—can be distinguished from work-related complaints (Table 8–1).

64

TABLE 8-1 Common Respiratory Complaints—Medical and Industrial

Category	Medical Conditions			Industrial Conditions		
	Diagnosis	Symptom/Finding	Example	Diagnosis	Symptom/Findings	Examples
Mucous membrane symptoms	A "cold" (upper respiratory infection) Allergic rhinitis	Cough, cold, watery eyes, runny nose, ±fever ±Related to season, area. Unrelated to work	Common cold Hay fever, animal dander, aspirin allergy	Mucous membrane irritation	Nose, throat ±eye irritation Fever, cold symptoms are uncommon Transient, related to specific job activity	Barbecue smoke Household ammonia Chlorinated pool Smoky "pub" L.A. freeway
Asthma (wheezing, airway obstruction)	Non-industrial	Unrelated to job; positive family or atopic history Seasonal ±RAST positive With or following infections History of tobacco or marijuana abuse	Endogenous asthma With allergic rhinitis Bronchitis Sinusitis Pneumonia Flu COPD	Occupational asthma	Clear-cut job-related history Produced by specific activity or specific exposure For aggravation of non-industrial asthma, see Chapter 9	Isocyanates Epoxies Polyurethanes Uncured "resins" Wood dusts Vegetable dust
Flu-like reaction	Flu	±Fever, myalgias, common	Influenza, viral	Flu-like reaction	Uncommon; worker will usually self-diagnose second time	Uncured epoxy or other resin systems used for coatings or in plastics manufacture, metal fume fever

± may or may not

Non-industrial and industrial conditions, causing similar symptoms, are compared in this table.

I. COUGH

A cough can be a way of clearing the throat of mucus, a sign of viral bronchitis or bacterial infection, a symptom of a tumor in the chest, a result of too many cigarettes, or a response to an airborne irritant. The patient may work with some sophisticated-sounding chemicals, but if he is sneezing and coughing (even if he is taking vitamin C and working out), the odds still favor a cold. If it looks like a cold and sounds like a cold — it is a cold.

II. HAY FEVER, ALLERGIC RHINITIS

Typically, a patient with watery eyes, runny nose, and a cough would have a "cold." Depending on circumstances, this could also be hay fever or allergic rhinitis. But, in hay fever or allergies, the mucous membranes look different — they are blue and swollen. Many complaints attributed to workplace substances are really allergic rhinitis — people like something to blame. The diagnosis is based on history, lack of relation to changes in work practices (i.e., no change in symptoms upon elimination of contact or relocation), physical findings and, most of all, common sense. A patient does not need a work preclusion, "avoid all chemicals," every time hay fever season rolls around. Besides, some people have "perennial allergic rhinitis" — blue and boggy mucous membranes all year. These symptoms will plague them, no matter what.

In Nothern California, allergies are prevalent. Nearly everyone complains of mucus, phlegm, chest heaviness, or throat tightening to one degree or another. For the past several years, which have been under drought conditions, symptoms have intensified and may last virtually year-round. The physician should not be misled by the patient's account that symptoms improve on vacation, since allergies benefit from geographic change. Besides, who does not feel better on vacation? While some persons may be greatly disturbed and annoyed by the symptoms of allergic rhinitis, invariably they will not be job-related.

III. CHEMICAL IRRITATION

How can symptoms of a cold or allergic rhinitis be distinguished from those of chemical irritation? Handily, because with chemical irritation, there is the following:

- A clear history of exposures to irritants
- Possible eye irritation
- Lack of fever and flu symptoms
- Clear relation of symptoms to a specific activity
- No symptoms away from that activity

Commonly experienced samples of mucous membrane irritants in everyday life are barbecue smoke, household ammonia, freshly chlorinated swimming-pool water, or, as in the old days, a smoke-filled room. All irritant symptoms resolve a few minutes after the exposure stops. In general, these would have disappeared by the time a person reaches the physician's office.

Drug use of all kinds must be included in the differential diagnosis of "toxic" problems. Eye irritation could certainly result from contact lens use or medications instilled into the eye and nasal irritation from street drugs or tobacco abuse.

IV. ASTHMA

Asthma is a common complaint. Asthma is a narrowing of the small passageways or bronchi and bronchioles leading to the air sacs of the lungs. For a diagnosis of asthma, objective evidence such as the presence of wheezing and pulmonary function test abnormalities is needed; a history of "asthma" is sometimes inaccurate.

Non-Work-Related Asthma

Non-work-related asthma is caused by the following:

- Unknown factors, sometimes genetic
- Bacterial infections
- Allergy to pollens, foods, sulfites, etc.
- Medication/drug reactions.

The precise cause of asthma is often unknown. Of the known causes, sometimes asthmatic symptoms of wheezing and chest tightness are part of the symptom constellation of allergic rhinitis or hay fever. Post-infectious asthma begins following an infection such as sinusitis, bronchitis, or pneumonia (pneumonitis). The "itis" in these conditions stands for inflammation, which may be due to bacterial infections. None of these conditions would ordinarily be caused by work factors.

Chemicals used as drugs, such as beta-blockers, may produce asthma, as may aspirin, food allergies, or sulfites. Sulfites may be present in foods, drugs, or workplace air.

Wheezing, the hallmark physical finding of asthma, may also result from street drugs or cigarette abuse or from a blockage in the airways, such as a tumor or foreign body. The person with chronic obstructive pulmonary disease caused by smoking too many cigarettes or street drug abuse, wheezes, and this could be confused with a case of "asthma." Wheezing is just one small part of their problem, which is really chronic obstructive pulmonary disease (COPD).

Occasionally, a person has big tonsils or a lesion producing upper airway obstruction. These patients can be cured with a tonsillectomy, and they are grateful! There are many more asthmatics than there are people with asthma from work, and they need their jobs.

Removal from work of a person with asthma can be a disaster for the patient. Some persons with non-industrial asthma, who work well "with" their asthma, have been removed from fine jobs, only to learn that their asthma was unrelated to work. Unfortunately, these persons retain their asthma but lose their jobs, their health insurance, and other benefits—at great sacrifice to themselves and their families. Also, once they are "on record" as having asthma, they may be unable to obtain new health insurance or another job.

Occupational Asthma

While most cases of asthma are not work-related, it is important to recognize those that are, as they are often preventable and curable. Typical causes include exposure to:

- Isocyanates and constituents of uncured plastics, e.g., epoxy systems
- Sulfites as chemicals or off-gassing
- Natural or synthetic resins or gums
- Organic dusts

The typical profile of a person with work-related asthma is one who repeatedly tries to work but has a recurrence of asthmatic symptoms each time he/she is exposed to the offending materials. In contrast to patients with non-work-related asthma, the history is specific. Asthma is triggered by specific circumstances at work. In general, most are fine when they are not in the workplace. Uncommonly, some do not revert to their baseline status. It has been said that some persons with pre-existing asthma may be more susceptible to occupational asthma. Thus, a pre-hire exam might preclude a person with a prior history of asthma from working with or near isocyanates. Aggravation of preexisting conditions is further discussed in the next chapter.

If occupational asthma is a strong consideration, it is important *not* to put the patient on long-term medication. It is the history of asthma in response to a specific substance, and lack of response outside of the workplace, that makes the diagnosis. If the asthmatic trigger is removed from the workplace, there is no need for long-term medication.

V. PULMONARY FUNCTION TESTS

Pulmonary function tests may assist diagnosis of pulmonary complaints. Baseline tests are valuable when obtained before someone starts a potentially problematic job. Measures of airway function document a baseline and record changes related to exposures, as in isocyanate-induced asthma. Another test, diffusion capacity of the lung, (DL) is an accurate measure of lung tissue integrity or damage and cannot be faked.

VI. LARYNGOSPASM, ANGIONEUROTIC EDEMA

A patient may develop a sensation of throat closing due to something at work. This type of reaction is rare but potentially serious if the worker develops laryngospasm or angioneurotic edema (allergic swelling of the throat). It is relatively easy to identify because symptoms occur in relation to a specific activity, substance, or location in the workplace, and not at any other time.

VII. FLU SYMPTOMS

Flu symptoms are relatively common. Most persons will have no more and no less than a viral syndrome, and some may have AIDS. Flu-like symptoms from exposures are rare but dramatic. Workers often make their own diagnoses when, after using uncured resin systems in plastics manufacture or coating, or certain metals with improper ventilation, feelings of flu, malaise, and impending doom develop. Invariably the worker will make his/her how diagnosis the second time they have a reaction to the same material.

Once a year, during maintenance, Jeff, a plant operator, welded carbonized steel. Last year he took time off from work at the conclusion of the welding assignment, thinking he had the flu. This year, he developed the same flu-like symptoms following a similar welding assignment. However, on this occasion Jeff realized that he did not have the flu but had a reaction to the fumes generated in the welding operation.

These cases are relatively straightforward and can be summarized as follows:

- Most complaints presented to a physician will not be work-related.
- Symptoms of chemical irritation require exposure to an irritant and are generally transient.
- Most asthma cases are non-industrial.
- Industrial asthma cases are frequently caused by isocyanates, sulfites, resins, or organic dusts.
- Industrial asthma cases give clear-cut histories.
- Industrial asthma cases should *not* be treated with long-term medication.

Aggravation of preexisting conditions, which is more of a gray area, is discussed in Chapter 9.

9

Diagnosing Significant Aggravation of Pre-Existing Conditions

KEY TOPICS

I.   Bronchitis
II. Asthma
III. Smoking

In the preceding chapter, a clear distinction was made between non-work-related conditions such as colds and flu and work-related conditions such as mucous membrane irritation from smoke and soot. Sometimes, the patients' presentation is not so clear-cut. There may be gray areas or instances where a person has a non-work-related condition, but it is aggravated by work. Some typical examples follow.

I. BRONCHITIS

Bronchitis can be a bacterial or viral infection, commonly occurring in winter. Ordinarily, even if bronchitis is "going around" the office, it is not considered to be an industrial condition. However, if there is a release of a gas, such as chlorine gas, and a worker develops prolonged bronchitis, even if infected, irritation by chlorine predisposed that person to developing bronchitis and/or prolonged it. But for the chlorine, that infection might not have occurred or would have been less severe.

70

II. ASTHMA

Attacks in asthma, a common non-industrial condition, may occur:

* randomly
* in response to specific substances
* because of emotional stresses

Inasmuch as attacks are random, some attacks may occur at work. The severity of attacks varies, but an "asthmatic" attack consists of 1) some degree of narrowing of the air passages, 2) a subjective feeling of being unable to move air in and out of the lungs, and 3) wheezing—high-pitched noises reflecting spasm of the airways.

To determine whether a given instance of asthma or period of exacerbated asthmatic attacks is significantly aggravated by the job, let us analyze a few cases.

An employee of a cleaning service used an industrial-strength tile cleaner on a bathtub in a small bathroom. She developed sustained and severe asthma —"status asthmaticus"—and was hospitalized. Eventually this attack subsided, but she did not return to her prior baseline status of having normal airway function in between asthmatic attacks.

Does this case manifest significant aggravation by an industrial factor? Ammonia, contained in the tile cleaner, being an irritant, seemed to precipitate or aggravate her asthmatic state. Other causes of asthma, such as an infection, were not found while she was in the hospital. Six months later, she had some residual disability in that there was some airway narrowing and breathing required more effort even when she was maximally medicated. Before that, she had been normal between attacks. This case would represent a significant aggravation by work-related factors of a non-industrial asthmatic condition. Now, if the same person were an office file clerk who developed a bronchitis which worsened the asthma and then did not return to her baseline status, would this be significant aggravation by a work-related factor? No, because there did not seem to be any identifiable work-related factors that would significantly aggravate the asthmatic condition as we have described it here. But, if the file clerk was being sexually harassed on the job, or had a mean-spirited or vituperative supervisor, and developed asthmatic attacks in increased frequency or sustained a constant level of airway narrowing, this could constitute a significant aggravation of a non-industrial condition.

III. SMOKING

In smokers who develop chronic obstructive pulmonary disease or lung damage, the question often arises, "How much injury would be apportionable to smoking, and how much to work-related factors?"

We will see in Chapter 13 the dose-response between cigarette-smoking and lung cancer; a similar response exists for cigarettes and airway and lung damage. In order to judge whether a cigarette-smoker has injury due to work-related factors over and above that from cigarettes, one needs to compare the dose-response between cigarette smoke intentionally inhaled and on-the-job airborne contaminants. In order to decide whether work contributed to pulmonary disability over and above that from smoking alone, it is necessary to show that the degree of disability occurred sooner or more severely than it would have if the work-related factor was not present. If, irrespective of a slight solvent odor in the workplace, the person who smokes four packs of cigarettes a day would have developed pulmonary obstruction of the same severity and at the same point in time, then it would not be considered a work-related aggravation.

The same type of reasoning can be used to analyze other kinds of conditions where an impairment is present.

To sum up,

- A pre-existing condition can be considered aggravated by workplace factors if, without the workplace factor, the disease would have been milder.
- Non-industrial asthma occurs at random and so may occur at work.
- However, just because an attack occurs *at* work, this does not necessarily mean it was caused or significantly aggravated *by* work.
- Non-industrial asthma can be aggravated by industrial factors if industrial factors such as airborne irritants, stress, or meanness significantly contributed to worsening of asthma.
- In a smoker (who already has a higher risk for many diseases from smoking), a workplace factor would have to significantly aggravate the condition over that which exists from smoking alone.

10

Common Sense Treatment and Work Restrictions

KEY TOPICS

I.   Common Sense Treatment
II. Common Sense Work Restrictions

I. COMMON SENSE TREATMENT

Since symptoms arising from chemical use on the job tend to be transient and abate when the person is removed from exposure, it is important not to commit a patient to long-term treatment. If possible, drugs should *not* be used. Brief symptoms may be replaced by drug side-effects, which obscure the problem. A patient will accept the explanation that you wish to observe his or her condition in relation to work and in the absence of drugs, especially if symptoms are short-lived. There is certainly a risk, such as with corticosteroids for asthma, opiate-like narcotics for pain, or benzodiazepines for anxiety, of treating a person for brief, trivial symptoms with drugs to which they may develop a long-term addiction.

> An agricultural worker developed nausea and vomiting. She thought she saw a crop sprayer leave the area as she was beginning her work shift. If she visited your office, the correct course of action would be to: a) obtain cholinesterase levels, b) hospitalize her for observation, c) reassure her, d) do a comprehensive history and physical, e) prescribe medication for pain and anxiety, or f) tell her it is nothing and send her back to work.

A combination of things were done in this case. First, cholinesterase levels were done and were normal. A comprehensive history and physical examination should have been done but was not. She should have been reassured when normal test results returned. When the woman continued to complain of nausea and vomiting, she should have been re-evaluated, but was not; instead, her symptoms were attributed to anxiety. She was told "it was nothing." When she continued to complain of feeling sick, she was treated with "big-gun" anti-anxiety and headache pain medication. Several months later, her doctors finally considered the possibility of pregnancy. Indeed, she was, but she had a miscarriage after taking all that medicine. Her doctor incorrectly assisted her in submitting a claim for work-related loss of the child and permanent disability—as an industrial claim, when the condition was one of non-industrial pregnancy.

This woman could have performed her usual and customary work, which did not entail chemical exposure, while pregnant. However, the use of medication in this case was of no benefit and was probably harmful.

Common sense treatment of chemically-related conditions would include the following:

Mucous membrane irritation	Reassurance Alter work practices Irrigation and topical treatment for eye irritation, if severe
Solvents	Reassurance Alter work practices Gloves, respiratory and skin protection, as appropriate
Organophosphates	Support vital functions Reassurance Specific antidotes, as medically appropriate
Industrial asthma	No long-term drug commitment Reassurance Avoid specific cause Observation
Highly reactive materials	Support vital functions Ensure adequate oxygenation Decontaminate Specific treatment, if available Reassure Avoid direct re-exposure

Emergency treatment is discussed in Chapters 14 and 15.

II. COMMON SENSE
WORK RESTRICTIONS

Let us say that one doctor is evaluating another—his partner in medical practice. From time to time over the last few months, he or she has been aware of a little tightness in the throat and a little heaviness in the chest. Would a work preclusion, "avoid all chemicals," and advice not to return to the workplace, make sense? Of course not! It would not make sense for the partner, or if the situation were reversed, if the examining doctor had to give up medical practice because of these symptoms!

Patients should be regarded in the same way. Perhaps the symptoms are allergic rhinitis, anxiety over family problems, or perhaps they are related to constituents of a new glue being used to build model planes or to fashion a prosthetic device. Common sense would dictate that the glue be used in a fume hood or replaced, not that the physician be rehabilitated into some other profession. It would certainly make sense to try to work with an employer to ensure that a patient returns to work safely with some relatively minor changes in work practices. If the patient is only too eager to leave the workplace, a healthy degree of suspicion should be aroused. Let us examine some work restrictions.

A new patient, a healthy young man 27 years of age, had trace blood in his urine (microscopic hematuria) on a routine physical. He is a maintenance mechanic in a plant with essentially no exposures other than cleaning his tools with solvents on a rag once a week, during which he wears heavy gloves.

The proper course of action in this case would be to: a) restrict him from his profession as a maintenance mechanic, b) restrict him from the use of solvents, c) find out whether or not he has ever had microscopic hematuria in past urinalyses, and d) find out what the solvent is.

The correct answers are c and d—obtain prior urinalyses, and obtain more information about the solvent. Unfortunately, this person did not work for years; he was "banned" from the workplace by several doctors. Someone had noted that in rodent studies, solvents can cause liver and kidney dysfunction, without bothering to inquire whether microscopic hematuria in people would be a reasonable expectation. In our opinion, he was a healthy individual with microscopic hematuria unrelated to the workplace. Chances are that microscopic hematuria pre-existed this job. While there was some *contact* with solvents, basically he had no *exposure.* Furthermore, the injury was not what one would expect from the solvent, in that microscopic hematuria would not be expected in a person who had no intoxication. The issue did not make sense; the worker suffered.

A woman complained of tightness in her chest and mucus in her throat, which she attributed to a pesticide from a nearby mosquito abatement project entering the air intake of her office building. The employer was very sympathetic and transferred her to a different building. She continued to complain. The entire firm was moved to a new location in yet a third building. Her complaints continued unabated. She eventually visited a specialist in "environmental illness" who banned her from buildings. She also submitted a claim for a back injury because she carried her portable typewriter to each building.

Some office workers smelled an odor of pesticide applied for carpet fleas when they arrived at work after a long weekend. Of several hundred employees, ten of them have been permanently "banned from indoor environments" as a result of this experience.

Are the buildings the problem? Of course not. That does not make sense. These persons carry their own baggage of chronic complaints with them but were not known to be living outdoors when last examined. They were only banned from indoor "work" environments. Does this case make sense? Of course not! Try this one:

A 25-year-old auto body spray painter spray-painted automobiles in a state-of-the-art spray paint booth. In addition, he used respiratory protection with fresh air supplied respirator and wore coveralls and gloves. He developed flu symptoms. A physician banned him from the workplace because of isocyanates in the spray paints.

Does this case make sense? No, it does not, because flu symptoms from isocyanates are, first, uncommon and, secondly, are virtually unheard-of without pulmonary findings. This man had no pulmonary findings or complaints. He did have fever and gastrointestinal symptoms, and it sounded very much like the gastrointestinal flu. At the least, in this case, he should have been allowed to return to work and then observed carefully for a recurrence of symptoms.

A physician should not impose a work restriction on a patient because the doctor does not understand what is going on. Any work restriction that is imposed must be justified by a *very* good reason. If the physician keeps an open mind, a patient can be managed and cared for expertly, even if all the facts are not clear, as described in the next chapter.

11

How to Diagnose and Manage the Patient with a Possible Work-Related Injury

KEY TOPICS

I. Diagnosis
II. Essentials of an Evaluation Report
III. Obtaining More Information

I. DIAGNOSIS

What does a family doctor need to know about his patients in order to decide whether a work-related chemical injury has occurred? Imagine for the moment a family doctor practicing in an industrial community. In a typical week, seven workers appear with the following complaints:

- Marv complained of fever, malaise or being tired—the flu.
- Chickie had a rash on his forearms.
- Harold complained of shortness of breath.
- Tim was concerned about decreased sexual function.
- Bob had no complaints but was there for a routine exam. Liver function tests, consisting of alkaline phosphatase and gamma glutamyl transpeptidase (GGT) values, were abnormal.
- Roberto and Louie both complained of asthma.

Briefly recapitulating, there are some key questions to be answered when evaluating each of these individuals—our Seven-Step Toxic Verification Test:

- What is the person's motivation for the visit?
- What is the diagnosis?

- Did the person sustain chemical exposure?
- Was the exposure of sufficient intensity to produce an injury?
- Is there evidence that an injury was produced?
- Is the injury likely to be a result of the exposure?
- Does the conclusion make sense?

The patients are all out there in the waiting room, and they need to know, and the union wants to know. Let us analyze Marv's case:

- Why is Marv here?
  Marv is here because he feels sick.
- What is his diagnosis?
  Flu or flu-like symptoms.
- Did he have a chemical exposure?
  He is a machinist and has some contact with cutting fluids and carbon, but his workplace is excellent, and he has virtually no exposure. No incidents had occurred.
  No exposure.
- Did the exposure produce an injury?
  No.
- Does he have an injury?
  Well, he has the "flu."
- Is his diagnosis related to an exposure at work?
  No.
- Does the conclusion make sense?
  Sure does. It looks like the flu, which is going around.

The "flu," a nickname for influenza, which is a viral infection, will be transient and should not recur with chemical re-exposure. However, if Marv is using carbonized steel, a new metal alloy, or is part of a group of workers in an area where an experimental process is being conducted or a new coating is being manufactured, and several workers complain of flu or similar symptoms such as rashes and fever, then a workplace cause warrants serious consideration. The physician and the company need to communicate and investigate.

If several workers who essentially do the same job and are working with the same compounds complain of the same symptoms, say, dermatitis of the hands when using a solvent without protection, the likelihood of a work-related reaction increases. On the other hand, two workers with rashes can also have nothing in common. For instance, if one of them developed dermatitis of the hands *after* wearing rubber gloves for an eight-hour shift, more likely than not he is having a reaction to the gloves.

His partner, on a Monday morning, could have a blistering reaction of poison oak from working in his garden over the weekend.

What does Chickie have?

- Why is Chickie here?
  Chickie is here because of a rash on his forearm.
- What is his diagnosis?
  Dermatitis.
- Did he have a chemical exposure?
  Yes. He is a janitorial worker, and he splashed a bucket full of "industrial strength" detergent on himself.
- Did the exposure produce an injury?
  Yup. The rash developed shortly following this incident.
- Does he have an injury?
  The rash is the injury; it is clearing up.
- Is his diagnosis related to an exposure at work?
  Yes. It is related to the detergent which caused it.
- Does the conclusion make sense?
  Yes. The detergent is caustic, and the skin reaction is predictable.

Chickie needs to be more careful and/or use a product that is not so irritating. Note that this is a compensable injury, but he had no lost time from work, temporary or permanent disability.

What about Harold?

Harold, who still smokes three packs of cigarettes a day, entered an environment where the odor of solvent was apparent.

- Why is Harold here?
  Harold complains of shortness of breath.
- What is his diagnosis?
  His diagnosis, based on his history of smoking three packs of cigarettes a day for 40 years, physical examination and pulmonary function tests, is chronic, obstructive pulmonary disease (COPD).
- Did he have a chemical exposure?
  No. He smelled an odor, but he has not had an exposure, per se. He did have some brief contact when he delivered a pallet of goods to the formulating area.
- Did the exposure produce an injury?
  No. The brief contact is not likely to be related to his pulmonary condition. However, to reassure Harold (and the doctor), pulmonary function tests as well as blood and urine tests that measure levels of solvent and solvent metabolites can be done.

- Does he have an injury?
  He has an injury—relatively mild COPD.
- Is his diagnosis related to an exposure at work?
  No. It is due to smoking.
- Does the conclusion make sense?
  Yes. The odor of solvent was insufficient to produce an injury and would not produce COPD. COPD, however, is the expected effect of his exposure to cigarettes.

Tim, a bachelor, was very shy and needed a lengthier visit and an exam before the doctor could determine the real reason for his visit. Tim was so worried about diseases and safe sex that he was unable to perform.

Ironically, another doctor prescribed an anti-anxiety medication that decreased sexual performance. Other patients taking histamine-blockers and blood-pressure-lowering drugs also became impotent. Let us apply the toxic injury verification test to Tim.

- Why is Tim here?
  Tim said he was here for allergy complaints but he really wanted to ask about potency problems.
- What is his diagnosis?
  Inability to develop or sustain an erection.
- Did he have a chemical exposure?
  He operates a chemical mixer for manufacture of floppy disks for computers. About a week before his problem started, he did splash some chemicals on his thigh.
- Did the exposure produce an injury?
  He did thoroughly wash it off right away. On the Material Safety Data Sheet, it said that this stuff caused reproduction problems in mice.

It took several visits and a rapport with the doctor to learn that Tim was so worried about catching AIDS that he was unable to perform. It took some time and patience for the doctor (and Tim) to realize that worry, not the chemical, was behind the problem.

- Does he have an injury?
  No. There is no physical injury.
- Is his diagnosis related to an exposure at work?
  No. He had a brief "exposure," which was coincidental to his symptoms.
- Does the conclusion make sense?
  Yes. There was a splash, and he had a symptom, but they were not related as cause-and-effect.

Bob felt well. He was there for his annual, routine exam, but had abnormal liver function tests.

Let us look at Bob.

- Why is Bob here?
  He is here for his annual checkup.
- What is his diagnosis?
  Abnormal liver function tests.
- Did he have a chemical exposure?
  He has been having too many martinis.
- Did the exposure produce an injury?
  He may be developing fatty liver or alcoholic hepatitis; we must get Bob to stop drinking.
- Doe he have an injury?
  He does have early liver injury, according to the physical exam (large liver) and the test results.
- Is his diagnosis related to an exposure at work?
  No. This cannot be blamed on his employment.
- Does the conclusion make sense?
  Yes.

What if Bob were a computer programmer or a carpenter?

- Why is Bob here?
  He is having his annual check-up.
- What is his diagnosis?
  Abnormal liver function tests, asymptomatic, probably a variation of normal.
- Did he have a chemical exposure?
  He takes no alcohol, no medication, no anabolic steroids, and no drugs; he is a "physical fitness nut." There were no exposures.
- Did the exposure produce an injury?
  There are no exposures that can be defined.
- Does he have an injury?
  There is no evidence of an infection such as hepatitis or other virus. Some people have perturbations in liver function without a disease.
- Is the injury related to an exposure at work?
  No.
- Does the conclusion make sense?
  Yes.

What if Bob were a laborer, gluing plastic pipe in bathrooms in an apartment house complex under construction?

- Why is Bob here?
  Bob is here for a check-up, but he had been feeling dizzy and tired.
- What is his diagnosis?
  Abnormal liver function tests; consider "chemical hepatitis."
- Did he have a chemical exposure?
  Yes. For the last five weeks he has been gluing plastic pipe all day long, eight hours a day, five days a week—he worked in small spaces, without ventilation, using no personal protection. He has been reprimanded twice for being a slob. He doesn't believe in showering, either.
- Did the exposure produce an injury?
  Yes. Bob seems to have excessive solvent absorbed through his skin, which can be injurious.
- Does he have an injury?
  Yes. Chemical hepatitis, mild, without jaundice.
- Is his diagnosis related to an exposure at work?
  Yes. Solvents have produced this type of effect in similar circumstances.
- Does the conclusion make sense?
  Yes. Bob was taken off work and improved within a few days. The other man who was gluing pipe, Antonio, had the same aberrations in liver function tests. The employer implemented new safe work rules so that no more cases occurred.

Liver function test changes are hardly ever due to the workplace. They occur in a certain number of normal individuals from prescription drug and/or alcohol use; rarely, if ever, from the workplace.

Roberto and Louie worked for the same contractor; both complained of asthma. Roberto was first to be examined.

- Why is Roberto here?
  Roberto complained of inability to breathe—asthma. He never had it before in his life.
- What is his diagnosis?
  He gave a history of asthma; on physical exam he is wheezing, and pulmonary function studies are consistent with asthma. He has asthma.
- Did he have a chemical exposure?
  Yes. He has been working with a "special paint" for the last three days without respiratory protection other than a paper dust mask.
- Did the exposure produce an injury?
  The label for the paint indicates that it is an "epoxy" paint. Some of the constituents have been associated with asthma.

- Does he have an injury?
  Yes. He has asthma.
- Is his diagnosis related to an exposure at work?
  Yes. It seems that unprotected work caused him to breathe in the fumes and develop industrial asthma.
- Does the conclusion make sense?
  Yes. The attack subsided; he no longer works with epoxies and has not had another attack.

Louie was a different story.

- Why is Louie here?
  He was having asthmatic attacks.
- What is his diagnosis?
  Asthma, from infancy.
- Did he have a chemical exposure?
  He was doing carpentry on the same construction site as Bob (the pipe-gluer) and Roberto (the painter). He had only been on the job three days, nailing boards in an area where no painting, pipe-gluing or other activities that could introduce airborne contaminants were occurring. However, he partied too much over the weekend and was now taking a lot of aspirin. Chemical sensitivity to aspirin is a possibility. He has polyps in his nose, which are associated with an allergy to aspirin.
- Did the exposure produce an injury?
  Not an on-the-job exposure.
- Does he have an injury?
  He has lifelong asthma plus probable aspirin sensitivity and nasal polyps.
- Is the diagnosis related to an exposure at work?
  His diagnosis is not related to work. However, if, on the job, he hit his hand with a hammer and took aspirin or anti-inflammatory drugs for that, which made his asthma worse, that attack would be caused by a work-related factor.
- Does the conclusion make sense?
  Yes. Louie has preexisting asthma with probable aspirin sensitivity. He should avoid aspirin.

Note that Louie did not have a chemical exposure. If one were to change the facts so that he had exposure to dust, the possibility of wood dust aggravating preexisting asthma would have to be considered.

II. ESSENTIALS OF AN EVALUATION REPORT

Sometimes a physician is asked to evaluate a patient to determine whether an illness or complaint is work-related and to write a report. Workers' Compensation regulations vary from state to state. Published guidelines for such evaluations may be obtained from agencies in your state, but all are structured more-or-less along the following lines.

- Reason for referral
- Issue at hand
- Employment history with employer at issue
- Prior and subsequent occupational history
- Medical history
- Physical examination
- Laboratory results
- Review of medical records
- Diagnosis
- Comments and conclusions
- Disability
  - Temporary
  - Permanent
- Causation and Apportionment
- Medical management—diagnosis and treatment
  - Past appropriate
  - Present appropriate
  - Future required
- Vocational retraining
- Any other specific questions or issues

The cases of Marv, Chickie, Harold, Tim, and Bob were fairly straight-forward. However, after examining a patient, the physician may realize the situation is complex. Chemicals with impossible-sounding names and strange symptoms are involved. More time and information are needed to evaluate and reach an opinion.

III. OBTAINING MORE INFORMATION

With issues concerning toxics, it is not always possible to give a definitive answer without more information, and sometimes it is problematic even after that.

Here is a list of some things to do or set in motion in order to obtain the facts about toxics. It is very important to get the facts, and get them straight.

The Job:

- Obtain Material Safety Data Sheets or labeling information.
- Understand what the substances are.
- Understand how the job is done.
- Read about and study the problem.
- Estimate potential exposures.

The Drugs:

- Re-evaluate the amount and effects of all drugs that the person is taking, including illicit drugs.
- Get a sense of drug use.
- Do drug screens if necessary.

The Lifestyle:

- Assess substance use.
- Assess household, recreational, and other off-work "exposures".
- Re-assess motives if nothing makes sense.

TABLE 11-1   What to Do and What Not to Do

Do Not Do or Set in Motion the Following Negatives	Be positive: Do the Following Instead
Do not assume that because a person has a job and symptoms or illness, they are related as cause and effect.	Use common sense; research the facts.
Do not reach hasty, wrong conclusions.	Take the time to think and research or refer the patient out.
Do not transmit your insecurity, anxiety, or ignorance to the patient.	Tell the patient that more information is required to reach an intelligent decision. Obtain it.
Do not give the patient unrealistic expectations of when you will know.	Investigate as quickly as you can and enlist the help of the patient, his family, and resources as appropriate.
Do not give the person a work restriction "Avoid All Chemicals" or "Toxic Exposures" — this is not a diagnosis.	A correct diagnosis will have a clear, specific and focused work preclusion.
Do not give a hasty work preclusion or restriction that costs your patient a job or livelihood.	Think of the implications your restriction will have on the person's life, try to enable him to conduct his job in a healthy way, which is possible most of the time. At least get a second opinion before the job is threatened.

Please be sure that the toxics area is one area in which it is very appropriate to tell your patient that you need more information. If you have neither the time, the interest, nor the inclination to handle this matter, whatever the reason, a referral elsewhere may be appropriate. Some practitioners really mean that they do not have the time or the inclination, but they do not verbalize it. Instead, by their words, actions, or inaction, their behavior has a negative impact on their patient and can give the patient a false impression of the true nature of their condition.

If the case is problematic, one's insecurity, anxiety, or ignorance should not be transmitted to the patient. Patients who are alarmed and frustrated have suggested that attention to the "Do Not" list on Table 11–1 can be of great benefit to them. Do not give them a hasty work restriction, which may affect the rest of their lives, until you have all the facts. Do not assume that because a person has a job and an illness, they are related as cause and effect.

Use your time efficiently and wisely to reasonably explain to the patient that more information is needed, and enlist their help in acquiring it. Investigate the situation as you have the time or inclination by communicating with the company, or refer this patient to someone who will. In the absence of facts, be reassuring and compassionate, and use common sense.

12

How to Identify Claims That Are Not Work-Related Injuries

KEY TOPICS

I. WORKERS COMPENSATION SYSTEM IS NOT DESIGNED FOR TOXIC INJURIES

Resources for work-related injuries are limited. Claims due to factors other than work related injuries will deprive needy workers. In this chapter, some common characteristics of falsified work-related and non-work-related claims are described in order that doctors, employers, and insurers can better manage them.

As discussed in Chapter 21, permanent disablement from toxic exposures are *very* unusual. The workers' compensation system was not devised for toxic injury. For example, a person overcome with carbon monoxide on the job, who loses consciousness, is hospitalized, recovers fully and returns to work is entitled to pay for lost work time and medical bills. There is no allotment for permanent disability, "damage" or "blood" money. It does not provide awards for potential hazards or because some-

87

thing happened — only for permanent injuries. This leads to "fear of chemical claims" and exaggerated symptoms *after* full recovery because people feel they are entitled to money because they were sick or at risk.

II. TEMPORARY SYMPTOMS THAT LAST FOREVER

Permanent Disability Claimed for a Temporary Illness

As previously noted, a person many have dizziness or headache from carbon monoxide exposure; full recovery is the rule. Yet, claims of permanent disability, brain damage, and all sorts of symptoms have been claimed by persons who had mild elevations in blood carbon monoxide — blood levels similar to those found in cigarette smokers. Most genuine cases of toxic exposures, as described in preceding chapters cause temporary symptoms, but exaggerated symptoms and claims of permanent disability are common.

Kitchen Sink

Frequently, legitimate exposure or temporary disability triggers an injury claim. However, in addition to the work-related disorder, every condition the person has had or will have gets thrown in as a work-related illness. This type of claim is a close cousin to the exaggerated symptoms described using the example of carbon monoxide exposure.

III. CLAIMS THAT DO NOT MAKE SENSE

Toxics claims, that are not due to the job, frequently do not make sense. If a case does not make sense, there is something wrong with it. A work preclusion "*avoid all chemicals*" or a diagnosis of "*toxic exposure*" raises suspicion at once. Who of us, in this chemical world, can avoid chemicals (see Chapter 2)? These terms suggest that the writer is uncertain of a diagnosis. Otherwise, there would be a preclusion to avoid something specific.

IV. MEDICAL ILLNESSES MASQUERADING AS TOXICS CLAIMS

People have illnesses, but few are actually connected with work activities. Sometimes an odd diagnosis is a case of inadvertent misdiagnosis. Sometimes a diagnosis of a work-related injury is intentional because a person lacks other health insurance. In such cases, claims examiners need to be watch dogs, insisting on specific diagnoses and/or work preclusions. A second opinion may be required.

V. DRUG AND ALCOHOL EFFECTS

Some claims of symptoms from workplace toxic exposures are really caused by the effects of legal or illegal drugs. Symptoms are often treated with more drugs, rather than with removal of the offending agent. In any patient, drugs must be considered as a cause of symptoms. Persons complaining of drug side-effects frequently "present" as a case of toxic injury. People take drugs for aches and pains from many causes—from assembly line work, exercise, home remodeling, and so forth. Anti-inflammatory drugs, tranquilizers, allergy medication all have side-effects. Common side effects are the following:

- Depression
- Dizziness
- Lethargy
- Fatigue
- Sleepiness
- Rashes
- Memory loss
- Heightened anxiety

Some patients manipulate their doctors into prescribing drugs under the guise of a work-related injury. In some cases there is no work-related injury but rather a ruse to obtain drugs. In other cases, when prescription records are analyzed, an addiction problem is obvious.

Review often reveals a long-standing drug problem pre-existing an alleged injury.

A person always had a drug abuse problem. As a teenager, she abused cocaine, heroin, and marijuana. As an adult, tranquilizers, valium, and prescription narcotics replaced them. A prescribing doctor finally realized that the patient's "aches and pains" were a ruse to obtain drugs. He wrote in his records "I am giving you this last prescription, don't come back to see me for three months". The very next day, this person had an on-the-job "back injury," lifting a sack of dog food.

VI. TERMINATIONS, LAY-OFFS, AND DISPUTES

Toxic claims sometimes proliferate during company lay-offs, terminations, or to make a point in employer/employee disputes. If a company's medical department never had a claim during fifty years of operations and has a hundred similar allegations following a lay-off, be suspicious. Termination physicals, before a plant closes, can discourage some of these claims.

VII. SOMETHING IS THERE

One of the most frequent types of claims is provoked because a chemical is present in the workplace. The worker comes in clutching a Material Safety Data Sheet. In such cases, although the presence of a chemical is incorrectly equated to a hazard, there is no indication that it has caused an illness.

VIII. CRAZY CLAIMS AND UNCONVENTIONAL PHILOSOPHIES

Some physicians believe that if a person has a job, it must have made him sick even though there is no evidence that work caused ill health. Ah, for the Protestant work ethic of yesteryear! Some "environmental illness philosophers" have capitalized on the toxic scare. "Buzz" words to watch out for in these claims include the following:

- Environmental illness "or allergy," clinical ecology
- Immune system damage
- Permanent sensitivities
- Psychic stress
- Brain damage, headaches, "loss of intellect"
- Immune disregulation

Some claims are submitted as class action lawsuits alleging mass illness, such as "immune disregulation." Costly, unconventional treatments are prescribed including desensitizing antigens, detoxification, germanium and "activated oxygen" tablets. Persons allege afflictions, such as permanent sensitivities to the environment, including allergies to diesel fumes, restaurants, newsprint, supermarkets, perfumes, and felt-tip pens.

Some self-proclaimed victims of environmental illness perceive themselves as chronic invalids and may join support groups of other victims of "environmental illness." Some claim intolerance to indoor environments or buildings, others are happy only on the beach.

The unwary should be alerted to complaints of brain damage, injury to the psyche, headaches, psychic stress, or loss of intellectual capacity. Typically, loss of intellectual capacity claims allege that a janitorial worker has lost the intellect of a physics professor.

When confronted by odd claims such as those described in this chapter, the examining physician, employer, or claims adjustor must:

- Reconsider the diagnostic criteria
- Review treatment
- Monitor drug costs
- Challenge odd diagnoses or treatments

13

Work and Cancer

Patients ask about carcinogens and need explanations. They read about them on fact sheets, labels, Material Safety Data Sheets, posted regulations, and union newsletters. The constant bombardment engenders a lot of fear because people do not know what this information means.

Before we consider the relation between work and cancer, what are the latest facts about cancer in the United States? The risk of dying of cancer is known: one out of four deaths in the United States is due to cancer. The breakdown of the largest groups of cancers, according to the American Cancer Society, is as follows:

- 24 percent lung
- 13 percent large bowel
- 9 percent breast
- 5 percent prostate
- 4 percent stomach

Tumors in various other organs constitute the remainder.

What are the known causes of these cancers? First and foremost, personal habits. Doll and Peto have estimated that personal habits account for most cancers—in fact, 98 percent of them. Smoking is responsible for one out of every six deaths—not only cancers but deaths from all causes. Tobacco use accounts for about 36 percent of cancer deaths, with alcohol another three percent (Table 13–1). Unfortunately, medicines and medical procedures now *cause* about one percent of deaths from cancer, while the total from occupational, pollution, and industrial products would account for one percent or less.

Doll and Peto's review of the agent or circumstance, along with the organ affected by the cancer, is separated into social and medical *causes*. The "social" or "life and lifestyle" cancer-causing factors are indicated in Table 13–2. Table 13–3 lists medical causes of cancer. All of the established causes of cancer in humans are listed. Some others that are suspect in humans are not listed, such as the relationship between diet and cancers of the breast or colon.

TABLE 13-1   Factors Associated with Cancer Causation[a]

Factor	Estimated Perfect
Tobacco	36
Diet	36
Infection	10?
Reproductive/sexual	7
Sunlight and radiation	5
Alcohol	3
Genetic	2
Medicines and medical procedures	1
Occupation, pollution, industrial products	<1

a. After Doll and Peto, 1981.

TABLE 13-2   **Established Human Carcinogens and Carcinogenic Circumstances from Life and Lifestyle (social)**

Agent or Circumstance	Cancer Location
Aflatoxin (fungal toxin on grains and nuts)	Liver
Alcoholic drinks	Mouth, pharynx, larynx, esophagus, liver
Chewing (betel, tobacco, lime)	Mouth
Overeating to obesity	Endometrium (lining of uterus), gallbladder
Sexual and reproductive/female:	
Late age first pregnancy	Breast
Zero or low parity (child-bearing)	Ovary
Promiscuity	Uterine cervix
Parasites:	
Schistosoma haematobium	Bladder
Clonorchis sinensis	Liver (cholangioma)
Tobacco smoking	Mouth, pharynx, larynx, lung, esophagus, bladder
UV light (sunlight)	Skin, lip
Virus (hepatitis B)	Liver (hepatoma)

Source: Peto and Doll, 1981, p. 1203.

I. EVERYDAY CARCINOGENS: SMOKE AND SOOT

In California there is a regulation that applies to cancer in firefighters (California Labor Code 3212.1). If a firefighter develops cancer in the course of his work, and has reasonably been exposed to a carcinogen known to produce that type of cancer, his cancer is presumed to be work-related. It is no surprise that smoke and soot are carcinogens because soot was the first occupational carcinogen. In the 1700s, Sir Percival Pott found that cancers of the scrotum developed in chimney sweeps who had poor hygiene and accumulated soot there. Smoke, soot, and excessive dust, whether on the skin or in the lung, are irritating. In addition, smoke and soot contain many carcinogens—individual substances that caused cancer in animals when administered as separate compounds.

Today smoke is still the number one carcinogen. When I was a medical student, everyone smoked, including the Dean of the medical school; but slowly that began to change. At first it was so slow that it took two decades *after* cigarette smoking was clearly found to cause lung cancer for an official public health pronouncement. In 1979, the Surgeon General of the United States finally described cigarette smoking as "the single most important preventable . . . factor contributing to death . . . in the United States" (Surgeon General of the United States, 1979). Cigarettes are still freely available, and tobacco growers receive government subsidies.

TABLE 13-3   Established Human Carcinogens and Carcinogenic Circumstances from Medical Causes

Drug, Hormone, or Treatment	Cancer Location
Contraceptive steroids (oral contraceptives)	Liver (hamartoma)
Cyclophosphamide	Bladder
Melphalan	Bone marrow (nonlymphocytic leukemia)
Azothioprine	Bone marrow, skin, hepatobiliary, mesenchymal
Chlorambucil	Bone marrow
Busulphan (Myeleran)	Bone marrow, carcinomas
Chlornaphazine	Bladder
Immunosuppressive drugs (as a group)	Reticuloendothelial system
Ionizing radiation	Bone marrow and probably all other sites
Estrogens	Endometrium
Diethylstilbestrol (DES)	Vaginal cancers in daughters
Phenacetin in analgesic mixtures	Kidney
Steroids, anabolic (oxymetholone)	Liver
Arsenic and arsenicals (Fowler's solution)	Skin
Some combined chemotherapies for lymphomas (including mechlorethamine, vincristine (Oncovin), procarabazine and prednisone [MOPP])	Reticuloendothelial system
Methoxsalen with ultraviolet A therapy (PUVA) for psoriasis	Skin
Treosulphan	Acute nonlymphocytic leukemia

Source: IARC 1987 (Supplement 7)
          Doll and Peto, 1981, p. 1203

Rats and mice are not needed to count tobacco-related deaths; the human database is enormous. There are 56 million current smokers in the United States alone, one out of every four people, and cigarette-smoking "officially" has been declared the nation's biggest drug abuse problem (Surgeon General of the United States, 1988).

Based on 1985 data, of the 56 million smokers, each year a population equivalent to a city the size of Denver—390,000—dies from causes directly attributable to cigarette-smoking (Figure 13-1). Most deaths are due to heart disease and lung cancer, additional causes being strokes, chronic lung diseases, and other cancers.

Despite the enormous negative health impact of smoking over decades, public health officials have not acted effectively to mitigate this problem. In fact, the first group to actually *do* anything about the difference in death rates between smokers and nonsmokers was the life insurance industry.

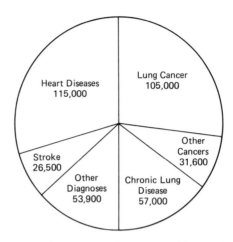

FIGURE 13-1. 390,000 deaths attributable to cigarette smoking (U.S., 1985)

They noticed that nonsmokers lived longer than smokers, who had higher death rates not only from respiratory cancers but also from other causes (see Figure 13-2). The life insurance industry charged smokers higher rates for life insurance than nonsmokers—a wise business decision.

Lung Cancers

According to the cancer statistics of the American Cancer Society, as depicted in Figures 13-3 and 13-4, between the years 1930 and 1985, in both men and women, cancer deaths for common cancers were either staying the same or going down, except for lung cancers. As can be seen from these figures, lung cancer deaths continue to rise dramatically.

Women certainly have "come a long way, baby." With increased smoking among women, lung cancer has surpassed breast cancer as the number one killer. The relationship between cigarette smoking and lung cancer in women can be seen in more detail on Figure 13-5. Just look at the four time periods between 1960 and 1986 indicated on the horizontal axis. The bars represent the age-standardized death rates per hundred-thousand women. The death rates for nonsmokers is represented in white, and that for smokers in black. It is evident that for nonsmokers there is a low background incidence of lung cancer which remains constant over this 26-year time period. But look at the black bars representing death rates in the female smokers—not only are the lung cancer death rates increasing,

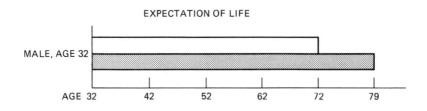

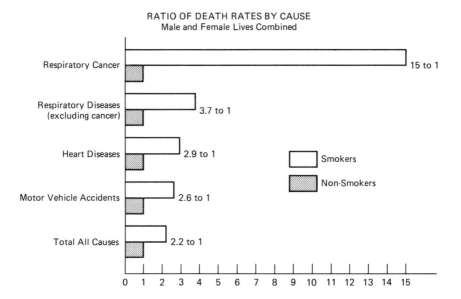

Based on "Mortality Differences Between Smokers and Non-Smokers", State Mutual Life Assurance Company. October, 1979

FIGURE 13-2.   Smokers and non-smokers

but they are *doubling* about every ten years. The skyrocketing risk of dying is true not only for lung cancer in women, but for a number of other smoking-related cancers, in both men and women, and will be discussed shortly.

First, it is worth examining the dose-response curve for cigarette smokers and lung cancer. One does not need rats or mice to look at the relationship between cigarette smoking and cancer deaths in humans (Figure 13–6). This figure indicates a positive dose-response between the number of cigarettes smoked (in billions) and number of persons dying of lung cancer (in hundreds of thousands).

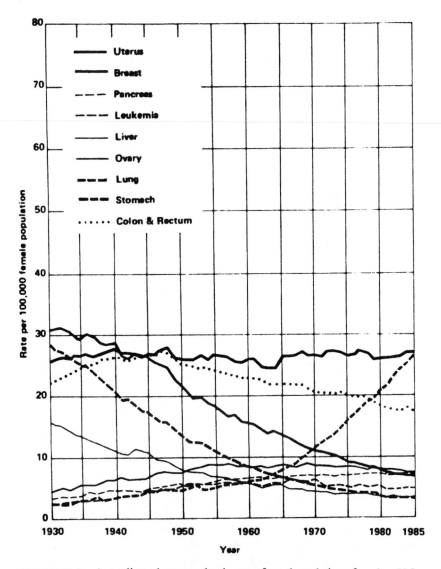

FIGURE 13-3. Age-adjusted cancer death rates for selected sites, females, U.S., 1930–1985. (Source: U.S. National Center for Health Statistics and U.S. Bureau of the Census. Rates are adjusted to the 1970 U.S. Census Population.) Reproduced from Silverberg, Boring and Squires, CA, 1990, p. 16.

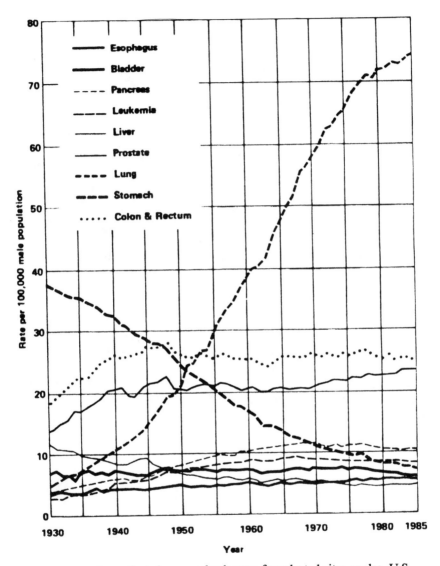

FIGURE 13-4. Age-adjusted cancer death rates for selected sites, males, U.S., 1930–1985. (Source: U.S. National Bureau for Health Statistics and U.S. Bureau of the Census. Rates are adjusted to the 1970 U.S. Census Population.) Reproduced from Silverberg, Boring and Squires, CA, 1990, p. 17.

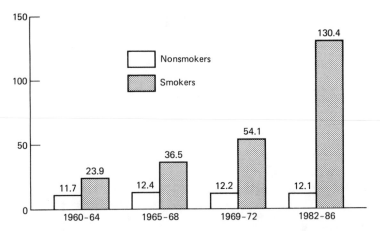

FIGURE 13-5. Female lung cancer death rates over time. Age standardized death rates per 100,000 women. (Source: Cancer Prevention Studies I and II. American Cancer Society.)

Risks of Smoking and Dying

Dying of lung cancer is not the only risk that smokers take. Two American Cancer Society sponsored large-scale prospective surveys of smoking and mortality, called the Cancer Prevention Studies (CPS) I and II, span a 56-year period. The study design allows comparison of the risk of dying in 1965 (CPS I) to the risk of dying in 1985 (CPS II) (Table 13–4). Note that

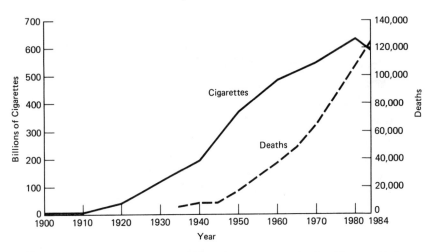

FIGURE 13-6. Cigarette consumption and lung cancer deaths. (Source: U.S. DHHS, PHS, Office on Smoking and Health.)

TABLE 13-4   Estimated Relative Risks, Persons 35 Years and Older

Cause of Death	Males 1965	Males 1985	Females 1965	Females 1985
Coronary heart disease (CHD)	1.8	1.9	1.4	1.8
Cerebral vascular accident (CVA)	1.4	2.2	1.2	1.8
Chronic obstructive pulmonary disease (COPD)	8.8	9.7	5.9	10.5
Cancer lip, mouth, pharynx	6.3	27.5	2.0	5.6
Cancer esophagus	3.6	7.6	1.9	10.3
Cancer pancreas	2.3	2.1	1.4	2.2
Cancer larynx	10.0	10.5	3.8	17.8
Cancer lung	11.4	22.4	2.7	11.9

Source: Cancer Prevention Studies I and II, American Cancer Society

this table expresses relative risks. The risk of a non-smoker dying of a condition is 1.0. The risk of dying of that condition if you smoke, relative to that of a non-smoker, is the smoker's risk. Note also that for lung cancer, the relative risk in men who smoked was 11 and 22 times greater than a non-smoker in 1965 and 1985. Although the present generation may be smoking about the same as their parents, the risk of dying of certain cancers is going up, and the risk of smoking and dying from cancers of the mouth, esophagus, larynx and lung is increasing dramatically. Just compare the risks between 1965 and 1985 for most cancers!

Carcinogens in Tobacco Smoke

It is not widely known that in tobacco smoke alone, more than 4,000 toxic chemical and radioactive compounds have been identified. Many of them will produce tumors when administered as individual compounds to animals, and thus, are carcinogens. Some carcinogens in tobacco smoke are the following:

- Benz(a)anthracene
- Benzo(b)fluoranthene
- Benzo(a)pyrene
- Cadmium and certain cadmium compounds
- Dibenz(a,h)acridine
- Dibenz(a,h)anthracene
- 7H-Dibenzo(c,g)carbazole
- Dibenz(a,j)acridine
- Dibenzo(a,h)pyrene
- Dibenzo(a,i)pyrene
- Indenol(1,2,3-cd)pyrene
- 2-Naphthylamine

- N-Nitrosodiethylamine
- N-Nitrosonornicotine
- N-Nitrosopiperidine
- N-Nitrosopyrrolidine
- 2,3,7,8-Tetrachloro-dibenzo-p-dioxin

Other toxins present in tobacco smoke are:

- Benzene
- Hydrocyanic acid
- Acrolein
- Ammonia
- Formaldehyde
- Acetaldehyde
- Carbon monoxide
- Nicotine

Many carcinogens and noxious substances are present in smoke, even from one cigarette, in levels much higher than amounts of corresponding compounds found in workplace air, situations driving headlines, or occurrences fueling "environmental lawsuits."

Smoking information should certainly be obtained as part of worker health screening. Today, more blue-collar than white-collar workers are smoking. Workers need to be advised about the risks of continuing to smoke. One marvels at how patients follow advice. Tell them to "avoid all chemicals" or stay home from work and they will do it yesterday. Tell them to "stop smoking" or "stop drinking" — the response is not so quick!

II. EVERYDAY CARCINOGENS: ALCOHOL

Alcohol is another popular carcinogen. Alcohol, consumed as a beverage, is responsible for three percent of all human cancer deaths (Doll and Peto 1981). As with cigarettes, there is a clear-cut dose-response between alcohol and related cancers. Some of the tumors attributed to alcohol are the following:

- Head and neck: mouth, hypopharynx, larynx, tongue, esophagus
- Gastrointestinal: liver, pancreas
- Prostate

III. EVERYDAY CARCINOGENIC RISKS

Bruce Ames and coworkers (1987) (Table 13–5) have compared cancer risks inherent in everyday activities by computations that use data from humans as well as data from animal experiments. In contrast to most tables, this one needs to be read from the bottom up.

TABLE 13-5   Relative Cancer Risks in Everyday Life Compared to Drinking Chlorinated Tap Water, 1 Liter a Day[a]

Possible Hazard Compared to Tap Water	Possible Hazard: HERP (%)[b]	Daily Human Exposure
14,000	114.0	Isoniazid pill (prophylactic dose, 300 mg)
4,700	4.7	Wine (250 mL)
2,800	2.8	Beer (12 oz.)
600	.6	Formaldehyde, 598 μg in conventional home air, 14-hour day
100	.1	Basil (1 gram dried leaf)
80	.08	Swimming pool, 1 hour (for child) chloroform 250 μg/L (ppb) average
70	.07	Brown mustard (5 grams)
30	.03	Comfrey herb tea, 1 cup, or one peanut butter sandwich
4	.004	Contaminated well water, trichloroethylene, 2800 μg (ppb)
3	.003	Bacon, cooked (100 grams)
1	.001	Tap water, 1 liter[a], chloroform, 83 μg (ppb)

a. 1 liter is approximately 1 quart.
b. HERP—Human Exposure Dose/Rodent Potency Dose. After Ames, Magaw & Gold, *Science* 236:271–280, 1987, Ranking Possible Carcinogenic Hazards.

Risks were computed and compared relative to drinking chlorinated tap water, about a quart a day, that contained a small amount of chloroform. According to the calculations, it was 30 times riskier to drink one cup of comfrey herb tea or eat a peanut butter sandwich than to drink a quart of water. In comparison, a child swimming in a chlorinated pool for one hour had 80 times the risk. Amazingly, one glass of beer was 2,800 times riskier, while wine was 4,700 times worse than water on a daily basis. Compared with water, a medication, isoniazid, that actually is lifesaving in tuberculosis, was 14,000 times riskier. As can be seen from Table 13–5, the risks as computed from animal experiments and human data, our perceptions, and how we use this information in conducting our daily lives, do not always agree!

IV. EVERYDAY CARCINOGENS: MEDICINE AND MEDICAL TREATMENTS

It is rather mind-boggling to learn that medicines and medical treatments are now responsible for causing one percent of all cancers (Table 13–3). Many successful cancer drug regimens are in themselves capable of causing cancers—if the person lives long enough. This result of cancer chemotherapy is an ironic trade-off—a person's life is saved from one cancer, only to enable him to live long enough to develop a second cancer as a side effect of the "cure." While one might be willing, in the desperate situation of dying from cancer, to make such a trade-off, drugs that people take daily —such as oral contraceptives, estrogens, anabolic steroids used by body builders, and analgesics—are also known causes of cancer. It should be pointed out that we are only now entering the third decade of oral contraceptive use, and therefore the real potential of contraceptive drugs to produce cancers may not yet be known.

V. WORK AND CANCER, CARCINOGENS ON THE JOB

Today, cancer cases caused by work are rare. In the aggregate, Doll and Peto found that occupational causes of cancer were one percent or less of all cancers. One hopes that workplace exposures have diminished sufficiently so that it will become even less.

It is hoped, as suggested by Nicholson (1988), that most of the causes of cancer in the workplace have been identified and that the factors leading to them have been corrected. Nicholson has compiled the dates of discovery of some of these associations, as shown in Table 13–6.

Currently, occupational cancers receive a lot of publicity. In some cases, such as cancers in heavily exposed asbestos workers, the cause-and-effect relationship is certain. Many other cases lack a cause-and-effect relationship between work and cancer. For example, one out of every four people in the United States gets cancer, and one out of every four persons smokes. Since some of these smokers work, some will develop cancer. Also, some cancers arise in people of working age, but it does not mean that work caused the cancer.

What are the facts? Carcinogens for humans, by occupation and cancer site, are listed in Table 13–7. This table is intended to be an overview, not

TABLE 13-6   A Chronology of Human Work-Related Cancer, Date Suspected, by Decade and Occupation

Before 1930	1930–1950	1950–1970	1970–1980
Benzene	Arsenic	4-Aminobiphenyl	Bis(chloromethyl)
Chromium	Asbestos	Boot, shoe	ether
compounds	Auramine	manufacture	Vinyl chloride
Nickel refining	Benzidine	Furniture	monomers
Soots, tars, oils	Hematite mining	manufacture	
	Isopropyl alcohol	Mustard gas	
	production	Rubber products	
	2-Naphthylamine	manufacture	

After Nicholson, 1988, p. 45.

TABLE 13-7   Occupational Circumstances and Cancers

Agent or Circumstance	Occupation	Cancer Site
Aromatic amines: 4-aminodiphenyl, benzidine, 2-naphthylamine	Dye Manufacturers, rubber workers, coal gas manufacture	Bladder
Arsenic	Copper and cobalt smelters, Arsenical pesticide manufacture, Some gold miners	Skin Lung
Asbestos	Asbestos miners, Asbestos textile manufacturers, Asbestos insulation workers, Certain shipyard workers	Lung Pleura (outside lung lining) (Probably esophagus, stomach, large bowel)
Benzene	Workers with glues and varnishes	Marrow, esp. erythroleukemia
Bischloromethyl ether	Ion-exchange resin manufacture	Lung
Cadmium	Cadmium workers	Prostate
Chromium	Manufacture of chromates from chrome ore, pigment manufacture	Lung
Isopropyl oil	Isopropyl alcohol manufacture	Nasal sinuses
Mustard Gas	Poison gas manufacture	Larynx Lung
Nickel	Nickel refiners	Nasal sinuses Lung
Polycyclic hydrocarbons in soot, tar, oil	Coal gas manufacturers, roofers, asphalters, aluminum refiners (exposure to certain tars and oils)	Skin Scrotum Lung
Radiation, ionizing	Uranium and other miners	Lung
Radiation, ionizing	Luminizers	Bone
Radiation, ionizing	Radiologists	Marrow, all sites
UV light	Farmers Seamen	Skin
Vinyl chloride	Polyvinyl chloride (PVC) manufacture	Liver angiosarcoma
Wood dust	Hardwood furniture manufacturers	Nasal sinuses
Unknown	Leather workers	Nasal sinuses

After Doll and Peto, p. 1238

detailed, since in many instances certain jobs appeared to be at risk, not all jobs in that industry. Also, there is some overlap on this list with our medical list, since arsenic appears again, mustard gas became useful in cancer chemotherapy, and the earlier radiologists and fluoroscopers developed cancers. In some instances, as in asbestos manufacture, large numbers of cases have been identified, while in others the listings are based on small excesses.

Looking at this material in the aggregate, how can the average physician intelligently assess a risk that a patient of his might sustain at his job? For one thing, if all the known causes of cancer in humans are taken into consideration, they more or less break down into six groups. These six groups of human carcinogens are easy to remember:

- Smoke, soot, excessive dust
- Highly reactive compounds (drugs, gases, chemicals)
- Hormones
- Chronic irritants
- Radiation
- Infectious agents

Smoke, Soot, and Excessive Dust

Remember that smoke and soot are carcinogens and that the first reported occupational cancer was cancer of the scrotum in chimney sweeps. If a worker has had a dusty, dirty job for decades, if he is a nonsmoker, and has been a chimney sweep without respiratory or skin protection, and has a cancer of the lung, skin or scrotum, then work-related cancer is a realistic possibility. If he once saw a fellow at work put out a fire in a wastebasket, had a desk job in an office building, and smoked four packs of cigarettes per day, then his cancer is not going to be industrial.

The lessons these cancers have taught us are that workers should avoid dirty conditions and use hygienic work practices. When work that involves smoke, soot, or excessive dust needs to be done, devices that purify air and protect skin must be used.

Highly Reactive Compounds

A second group of human carcinogens consists of compounds that are very chemically reactive. These are the anticancer drugs, immunosuppressive drugs, and poison gases like mustard gas—from which, incidentally, many anticancer drugs were developed.

In my opinion, uncured plastics and other highly reactive chemicals are potentially in this group. It is hard to predict 20 years into the future, but let us examine a prediction that was made in 1967. When Dustin Hoffman was "The Graduate," he was given just one word of advice: "Plastics." Simultaneously, the dramatic first cases of angiosarcomas of the liver—due to vinyl chloride monomer—were appearing. Plastics include materials such as acrylics, polyurethanes and epoxies, stable in a cured, hardened state; but fumes generated during manufacture, curing, burning, or liquid splashes, may be highly toxic.

What this means to workers is that highly reactive materials should be handled in enclosed processes and/or with the use of robotics, so that people need not be exposed to them.

Hormones

The same admonitions would apply to substances that are hormonally active, whether it be during their manufacture or use in animal husbandry.

Chronic Irritants

Chronic irritants are those substances that can stir up chronic inflammation. The compounds themselves do not necessarily have to be irritating; in fact, many of them, like asbestos or other fibrous materials, are inert. A fiber just sits there, in the lung, doing nothing. But when fibers enter and lodge in the body, the inflammatory cells—the body's security force—perceive these fibers and inert materials as being foreign to it. Armies of cells, often capable of laying down fibrous (scar) tissue, are stirred up on a chronic basis. This chronic inflammation may eventually result in a malignancy at the site where the fiber lodged. It is something like office politics.

Sunlight and Radiation

Radiation as a cause of cancer has long been known—since uranium watch-dial painters made fine points of their brushes with their tongues; and the early, uncontrolled high-dose uses of fluoroscopy and X-rays. Early radiologists were among the first radiation cancer victims. Fluoroscopy machines were even located in children's shoe stores to X-ray their feet and check the fit of the shoe.

Infectious Agents

Cancers are related to viral infections such as EB virus and AIDS, and viruses are suspected as being causal in many tumors of the blood and bone marrow, although as yet unproved. That is why infections are estimated to cause a minimum of 10 percent of cancers. While cancers from infectious agents as a result of on-the-job exposures are uncommon, AIDS-related cancers will increase in members of the health profession who acquire occupational AIDS, or in a worker who acquires AIDS during treatment arising from a work-related injury, e.g., blood transfusion for a fractured femur (thigh bone).

VI. DECIDING JOB-RELATEDNESS OF CANCER

Since one-fourth of all of us will, sadly, die of cancer, physicians will be called upon to decide, in a particular patient, whether a cancer is job-related. The best reference sources to assist in such a decision are the publications of the International Agency for Research on Cancer (IARC).

IARC consists of groups of scientists who meet periodically to review and update information on humans and animals with respect to the cancer-causing abilities of various types of chemicals. The lists of IARC carcinogens as updated in 1987 have been incorporated into the material in this Chapter (Tables 13-2, 13-3 and 13-7). When a job task or industry is listed as a human carcinogen in IARC, the decision for a patient becomes one of deciding whether a significant work-related exposure occurred. When there is no information that a compound causes cancer in humans, I find that, for a given patient, the results of animal tests need to be interpreted with caution when predicting what effect a substance will have in man.

VII. CONCERN FOR WORKERS AND WORKERS' CANCER CONCERNS

On a practical level, workers are nervous about working with or near "carcinogens." A company has to be aware of this and needs to show concern for its workers. The workers in turn need to take care to employ a high standard of hygiene on the job. One may assume that anything foreign to us is potentially toxic, and, therefore, the worker must enjoy his/her work while limiting potential exposures to the material used. The way to prevent workplace cancers is to *anticipate* and *prevent* the exposure. Reducing exposures will be emphazied in Chapter 16.

REFERENCES

Ames, B. N., R. Magaw, and L. S. Gold. 1987. Ranking Possible Carcinogenic Hazards. *Science* 236:271–280.

Doll, Richard and Richard Peto. 1981. *The Causes of Cancer.* Oxford University Press.

Silverberg, Edwin, Catherine Boring and Teresa Squires. 1990. Cancer Statistics, 1990. CA-A Cancer Journal for Clinicians 40:9–26.

Pott, P. 1975. Chirurgical Observations Relative to the Cataract, Polypus of the Nose, the Cancer of the Scrotum, the Different Kinds of Ruptures and the Modification of the Toes and Feet. London. (As cited in Doll and Peto.)

U.S. Department of Health, Education, and Welfare. 1979. Smoking and Health. A Report of the Surgeon General. U.S. Department of Health, Education, and Welfare, Public Health Service, Office of the Assistant Secretary for Health, DHEW Publication No. (PHS) 79-50066.

U.S. Department of Health and Human Services. 1989. Reducing the Health Consequences of Smoking. 25 Years of Progress. A Report of the Surgeon General. U.S. Department of Health and Human Services, Public Health Service, Centers for Disease Control, Center for Chronic Disease Prevention and Health Promotion, Office on Smoking and Health, Rockville, MD.

U.S. Department of Health and Human Services. 1988. The Health Consequences of Smoking, Nicotine Addiction. A Report of the Surgeon General. U.S. Department of Health and Human Services, Public Health Service, Office on Smoking and Health. DHHS Publication No. (CDC) 88-8406

Cancer Prevention Studies I and II, American Cancer Society. Atlanta, GA. Summarized in the U.S. Department of Health and Human Services, 1989.

World Health Organization International Agency for Research on Cancer. IARC Monographs on the Evaluation of Carcinogenic Risks to Humans. 1987 Supplement 7. Lyon, France.

Nicholson, W. J. 1988. IARC Evaluations in the Light of Limitations of Human Epidemiologic Data. Ann. NY Acad. Sci. 534:44–54.

Part 3

Advice on Effective Toxics Management

14

Advice for Employers and Insurers on Managing Exposures and Toxic Emergencies

A wise company is a vital part of the community on a daily basis. It also must plan for emergencies and catastrophes.

I. BE A GOOD NEIGHBOR

All companies should be good neighbors, but this is especially important for one that processes potentially toxic substances. As part of the community, personnel should be active in local organizations, schools and good works. Good public relations goes a long way toward counteracting toxic misconceptions, and facts can turn foes into friends. Do not wait for an army of adversaries to shut you down; be proactive and show the community your concerns and safeguards for them. Give plant tours to community groups. Sponsor toxics education programs at town meetings and schools. Use the media, such as newspapers, radio, and television, in a positive way that educates the public about company activities.

II. EDUCATE PHYSICIANS AND EMERGENCY RESPONDERS

Perhaps more than any other group, health care providers need to be educated.

> A chemical manufacturer asked a nearby hospital-based medical group to provide a medical surveillance program and emergency care for its employees; the chemicals had long names. The company was told, in no uncertain terms, not to bring the employees to the hospital for routine exams and *never* to bring them there in the event of a chemical accident.

The products were a bit unusual, even to a toxicologist, but the ingredients are used in children's toys. This attitude, from a health care provider, is a frightening reflection of the hostility that has developed against "toxics."

No one is *for* "toxics" but the hospital's attitude, while ignorant and uninformed, is not unique. This antagonism must be borne in mind while toxic substance management programs are developed—education of workers, communities, physicians, and the press begins on a hostile foundation. A new foundation is needed with a better mix of facts and common sense. Companies must keep health professionals who provide care—physicians, nurses, and emergency responders—informed.

Physician Education

A catastrophe requires community-wide preparation—be prepared. Ensure that physicians and hospitals are able to provide care in the event of a disaster; outside experts, as well as the company's insurer, may assist in the development of disaster plans. Tours for physicians, emergency responders, and hospital staffs will familiarize them with the substances used in the plant as well as acquaint them with the types of accidents that may be anticipated. Tours also introduce the medical community and plant personnel to each other thereby opening useful lines of communication. This ensures better medical care in the future when workers develop routine and uncommon medical problems.

The legitimate toxics area is complicated enough but misdiagnoses and wrong opinions make it much more confusing. Unless educated about a specific workplace, well-meaning physicians may incorrectly attribute a person's complaints to workplace exposures. The worker comes in, tells the physician he has been exposed to chemicals and, irrespective of the facts, a diagnosis of "chemical exposure" is made. The patient is usually advised to "avoid work with chemicals," a recommendation that is as impractical as it is wrong. As emphasized in Chapters 2 and 12, it is not possible to avoid all chemicals. If a worker is reacting to something in the workplace, it is something specific, not the Universe. This doctor needs education (and should read Chapters 2 – 11). He will be less likely to err if he is familiar with the company and problems that arise there; he needs facts about the workplace and substances used. Freed of toxic misconceptions, a physician can concentrate on real issues that affect his patient.

How a company selects and works with primary care providers in occupational health programs is discussed in Chapter 23.

Emergency Responder Education

Specific emergency providers, including plant personnel, local hazard response teams, emergency medical technicians (EMTs), fire fighters, ambulance and rescue crews also need working knowledge of potential crises. Emergency kits with specific antidotes, respirators, and protective clothing should be on-site at various locations inside and outside the plant and also provided to emergency responders at company expense if necessary.

III. PREPARE EMERGENCY PROTOCOLS AS A JOINT EFFORT

All accidents cannot be prevented, but advance planning does minimize injury to workers and the environment *after* an incident has occurred. The role of a company following an accident should be to:

- Keep management of the emergency simple
- Act as coordinator between various concerned groups
- Communicate with trained emergency responders and local health care providers

- Ensure appropriate treatment and decontamination procedures are followed
- Investigate
- Care about minimizing injury, confusion, and stress
- Communicate
- Correct problems

Have a Simple Plan of Emergency Procedures
On-Site, En Route, And At The Hospital

Two a.m. on a Monday with a skeleton crew was not the time for the other guy on the shift, who was overworked and overtired and also injured, to figure out what to do to save the live of his shift partner. There needs to be a clear plan of what to do on each shift if a predictable emergency arises and a general plan of procedure for the unpredictable. While some regulations dictate emergency and hazard plans, only the simple ones, implemented by people who understand them, will work.

All personnel should be educated but for each shift there should be a designated leader in case of emergencies (do not pick the fellow who faints every time he sees blood). A written plan should be a simple, single-page protocol for each type of emergency — something that a person at any educational level can understand and follow in a state of panic. Be mindful that any worker, even an articulate one, may not read, so the plan should contain pictures and/or color-coding, or the company should teach its workers to read.

Select appropriate locations for on-site first aid. Personnel should be clear on where to conduct toxics first aid. It makes no sense to have extensive decontamination going on downwind from a release if everyone downwind will be in contact with the cloud. Likewise, emergency equipment must be at locations convenient for treatment irrespective of wind direction.

Maintain emergency equipment in working order. An emergency plan should take into account life support as well as chemical decontamination procedures. All equipment needs to be maintained in working order and checked periodically — the simpler the better. It makes no sense to have the fundamentals of decontamination — emergency eye washes and showers, the mainstay of skin decontamination — if there is no running water. The whole plan must make sense. Adequate amounts of clean water at the right pressure, for use in an emergency situation, must be available.

Emergency responders should have the following:

- Appropriate protective clothing for themselves
- Material Safety Data Sheets (MSDS, see Fig. 19-2)
- Plans of procedure for your company
- Specific antidotes, if available

Ensure availability of appropriate equipment for decontamination and treatment. Workers must be trained to render aid to another individual without contaminating themselves. If one individual is drenched by a valve rupture or reactor release, the second should quickly suit-up him or herself prior to rendering aid. Then, all contaminated clothing (everything, including underwear—there is no room for modesty) must be removed, double bagged and appropriately reserved or disposed by procedures that have been decided previously. Decontaminate the person by a thorough washing and/or administering specific antidotes.

Implement emergency procedures to minimize injury. Injury, both physical and emotional, can be minimized by attention to the following:

- Support vital functions
- Administer specific antidotes
- Decontaminate promptly
- Reserve samples for biologic monitoring
- Reserve samples for area monitoring
- Periodically re-evaluate
- Care and show concern
- Communicate
- Reassure

Periodically re-evaluate. The patient must be thoroughly examined and re-evaluated in the event that the initial incident was perceived in error. For example, a lab technician who works with cyanide may be found unconscious so that a coworker fears a cyanide overdose. This is certainly possible but so is a heart attack or a stroke. From a medical and toxic standpoint, the injured worker should be thoroughly and comprehensively assessed with nothing overlooked. In this case example, personnel must be able to distinguish cyanide poisoning from other possibilities.

Remember the patient is a person, so frequently reassure. Generally, the exposed individual will benefit much more from human kindness and concern from the company and health care providers than anything that medical science has to offer.

Reserve Specimens And Develop Objective Data to Assess Exposures

Blood, urine, and skin wipe samples and other specimens, as appropriate (such as vomit), should be obtained from the injured worker and also from coworkers and medical attendants involved in the incident. While rarely is one person injured while assisting another, it has happened. Specimens from good samaritans will find their best use by reassuring them that they were not exposed.

Results of chemical analyses are rarely available on an urgent basis so treatment and management must proceed based on likely probabilities. Later the results of lab tests can generally be used to reassure the parties that little or no exposure occurred and also to plan further follow-up. (See Chapter 20.)

Objective tests provide valuable information. In my experience, the more objective data that are available, the sooner it is taken, the sooner the results are conveyed to the injured person—the more reassuring it is.

For example, Mike was drenched with a chlorinated solvent but due to quick action, very little was absorbed by inhalation or through the skin. Blood and urine samples showed that relatively little was absorbed and also on repeat sampling, established that the substance was gone from his body. He and his family were very much relieved.

Also, if later there is a difference of opinion over what happened and how serious the exposure was, measurement by a good lab can provide a proper set of facts. More often than not, a worker will be reassured to learn that his exposure was trivial, "Mr. Brown, you were exposed to an amount of solvent equivalent to one-tenth of a glass of beer." It is the worker who wonders months or years later to what he was exposed, and to how much, because "nobody ever told him anything," who becomes the problematic one—and often the plaintiff in a lawsuit.

IV. LISTEN AND COMMUNICATE

The injured worker, his coworkers, emergency responders, good samaritans, and/or distraught neighbors are all human. People can swallow pills by the handful and not worry, but the idea of even a few grains of a toxic substance being released is very upsetting to everyone. The company must do the right thing—show genuine concern, investigate the facts, keep open lines of communication and importantly, *listen* and *absorb* what people are saying. People are human and can upset easily, and it may get worse. The idea of "toxics" makes people angry.

If there has been a release or spill in the neighborhood, it is imperative that immediately a store-front, a van, or a field office be opened to function as a resource and provide a visible line of communication between the company and the community. This field office will play an essential part in shaping future events. Handled correctly, your representatives will coordinate with various concerned groups. People may have concerns and questions, may be genuinely afraid, they may wish to have blood tests done or have measurements taken in and about their homes.

The company should staff this office with their best and most personable people — people who are honest, calm, and reassuring. If your firm has no honest, calm, reassuring, or personable people, do the best you can. Hire a public relations firm or recruit some recent, happy retirees from the golf course for special action. There are also private consulting firms that can assist on an urgent basis.

You may or may not ever know exactly what happened, but it is so important to show people that you are doing your best to find out.

V. AFTERWARD, INVESTIGATE

What happened or what was released is not always immediately apparent. In fact, it usually is not. There is no shame in not having all the facts available in the immediate time period provided it is not perceived as a cover-up.

A checklist of information that is needed should be used and the gaps filled in. The incident report should contain information that helps physicians decide whether an individual has had a significant exposure. The worker, his family, the union, the insurer, and coworkers should know that an investigation is in progress. If appropriate, involve them in it, keeping them abreast of your progress.

Put yourself in a worker's place. The only way a worker will feel your concern is if it is genuine. It is important to care about each and every worker and take every incident, no matter how trivial it may seem at the time, seriously. People who feel they have been poisoned, get angry.

Even after the best of investigations, there may be some areas of uncertainty, so be it. It is not always possible to have a perfect answer to everything, but a sincere and open approach can go a long way in alleviating anxiety.

VI. CONVEY ONGOING CONCERN

Discuss corrective measures that have been implemented to prevent a recurrence of the incident. Allow time for a worker to return to his job, but, make every effort to get him back to work as soon as possible. The longer he is off, the less likely it will be that he will return (Chapter 23).

Do not make a big fuss about a worker in the first week or two and then ignore him. Build in an awareness or concern on the part of the company so even if management changes, the company will continue to follow him and convey ongoing concern. The respect that a company demonstrates for its workers will be repaid by the work force. A work force that feels that the employer cares about each and every one of them is likely to be happier and more productive with fewer accidents than one at a company where there is a collective feeling that human resources are expendable.

VII. LEARN FROM MISTAKES AND PUBLICIZE IT

It is a very good idea—as soon as the emergency is over and it is more or less clear what has happened or at least the questions are clear—to hold educational meetings for concerned individuals be they workers or members of the community. Explain what happened, explain why, if it is known. Discuss safeguards that have been implemented to prevent a recurrence. Provide an educational component on what the substance or substances are and how it may interact with people. If appropriate, include illustrations of everyday encounters with these substances as well as information on the dose response in rodents in comparison to human beings.

In sum, when an untoward event occurs, treat injured workers and the community as you and your family would wish to be treated. Show concern and kindness. Keep them informed. Be considerate and listen to what they have to say. Let an injured worker and the community feel that everything possible has been done for them—and do it.

15

A Checklist for Physicians, Nurses and Emergency Responders Managing Toxic Exposures

KEY TOPICS

I. When All is Calm and Bright
   - Items Health Care Providers Should Learn About
   - Items Health Providers Should Keep on Hand
   - Health Professionals Should Joint Venture Emergency Response Plans With and For Local Businesses

II. In An Emergency
   - Observe Treatment Priorities

III. Afterwards

This chapter provides a checklist of what health care providers need to do before an emergency happens, during a spill, an emergency, or a catastrophic situation and afterwards. These lists can be applied to other workplace injuries besides toxics.

I. WHEN ALL IS CALM AND BRIGHT

Health Care Providers should learn about:

- Local work activities
- Hazard and safety plans of local businesses
- Potential hazardous events that may arise
- Patients' or workers' jobs and job descriptions
- Chemical substances in local use

- Effects of chemicals in humans
- The Seven Step Toxics Verification Test

Health Care Providers should obtain and keep on hand:

- Material Safety Data Sheets from local workplaces
- Workplace hazard and safety plans
- Specific antidotes
- Procedures for decontamination
- Equipment for decontamination
- Several levels of protective equipment

Health professionals should joint venture emergency response plans with and for local businesses that include:

- Plant and workplace tours to educate health care providers about jobs and processes
- Joint planning with local companies for emergency situations
- Familiarity with company, industrial hygiene, health and safety staffs and, if available, company nursing and medical staffs
- Designated locations in plants and hospitals for chemical decontamination
- Designated procedures for disposal of contaminated items
- Case management toxicology training for personnel on various shifts
- Since during emergencies there is usually great confusion, plans should be simple, practical, and effective

II. IN AN EMERGENCY

The following treatment priorities should be observed:

- If necessary, don appropriate disposable protective clothing
- Support vital functions and stabilize the patient
- Decontaminate the patient by removing all clothing and cleaning the skin thoroughly
- Administer specific antidotes, if any
- Remember the patient is a human being; reassure frequently
- Reserve specimens. Draw plasma, serum, whole blood; obtain urine, refrigerate and/or freeze
- Obtain skin wipes and reserve
- Administer large doses of compassion
- Keep the patient and family informed
- Communicate with the company
- Educate yourself
- Avoid using drugs

Unless specifically indicated, avoid using medication, especially drugs that affect the nervous system or liver because drug effects mimic exposure effects. Accordingly, it may then be impossible to determine whether a person is suffering the effect of an exposure or a side-effect of the drug that he was given following an incident.

Emergencies typically occur at inconvenient times, such as in the middle of the night when skeleton crews are working—the release at Chernobyl, for example. The cause of an incident and the chemicals involved, are not always immediately apparent. The incident may need to be investigated and all information may not be available for medical personnel. By reserving specimens, later, when it is clear what has happened, biologic samples such as blood, urine and skin wipes can be sent for chemical analyses.

III. AFTERWARDS

After the emergency, physicians and company personnel should:

- Benefit from company fact finding
- Research the problem
- Communicate with the patient
- Obtain data
- Analyze biologic or area samples
- Educate the patient, the company, and interested parties in a spirit of common concern
- Synthesize information; provide answers if possible
- If all the facts are not available, say so
- Reassure the patient
- Decide whether expert advice or a second opinion is warranted
- Follow-up the findings with the patient
- Dispense continued reassurance

Remember if a patient has made it to the hospital and he/she is continuing to improve, it is likely that they will continue to do so. Afterwards, the health care team should learn from the experience and decide how they could have done better and implement those changes for "next time."

If uncertainty remains, the sooner the patient can be seen by a specialist who can provide reassurance, the better it will be for the peace of mind of all concerned, especially the patient. The feelings of persons who consider themselves to be victims of "toxic exposure" were described in Chapter 14 and are discussed in the next chapter. Reassurance and emotional support have a huge influence on recovery.

16

Advice to Employers on Reducing Exposures

KEY TOPICS

   I. Identify Chemical Inventory and Potential Toxic Exposures
  II. Prioritize Hot Spots
 III. Cost-Effective Solutions
   • Reduce Emissions and Wastes by Engineering Changes
   • Ensure Good Ventilation
   • Avoid Odors and Tight Buildings
   • Provide Protective Respiratory Equipment
   • Provide Skin Protection
   • Show Workers Respect

  IV. Design New Facilities Correctly
   V. Learn and Communicate

I. IDENTIFY CHEMICAL INVENTORY AND POTENTIAL TOXIC EXPOSURES

As the marketplace shrinks, the businessmans' worries escalate, toxics being only one of many concerns. How, in a sluggish economy, with cash flow problems, outmoded plants, and no resources for capital expenditures, can exposures be reduced?

A businessman needs to know what chemicals and wastes he has on hand and how much. Next, the chemicals that people are exposed to need to be identified. Then, those which may cause or have caused trouble with people should be determined.

Material Safety Data Sheets (MSDS) will be available for incoming raw materials and outgoing finished products as required by law. These may be intrinsically hazardous materials, such as acids, explosives, or a hazardous material if that is used improperly (like lead solder). Others, by their reactive nature, provoke symptoms in unprotected people — such as toluene di-isocyanate (TDI) or uncured components of epoxies when inhaled.

> In one workplace where workers are required to mix TDI with other components in an open tank, the average worker stays at the job for six weeks and then leaves clutching his chest because of respiratory distress due to asthma.

This property of TDI in humans is well-known — at least 10 percent of persons inhaling it will develop authentic industrial asthma. What are the options here? There are at least four:

1. Enclose the process so that nothing reaches the workers' breathing zone; aim for zero exposure.
2. Exhaust the process in a ventilated booth, and protect the worker by giving him a respirator supplied with fresh air.
3. Substitute something else besides TDI that does the job but not on people.
4. Enclose the process and use robotics.

No doubt there are other options.

Information is not necessarily available on mixtures, intermediate process streams, off-gas and curing fumes, by-products and wastes, but these are substances workers get into toxic trouble with. Now, these need to be identified. Chemists and the production staff will identify constituents of some streams and wastes; trade associations may provide information on others and provide information on health concerns common to certain processes or industries. Some trade associations have evaluated the toxicology of complex streams and mixtures, though not always in ways that are relevant to workers; American industry has financed a rat-mouse heaven of staggering proportion. The important thing is not to deny that your industry has accidents or health problems but to take a proactive role, learn from them and correct them.

II. PRIORITIZE HOT SPOTS

Identify and prioritize locations plagued by the following:

- Spills
- Frequent upsets

- Fugitive emissions
- Maintenance problems
- Frequent breakdowns
- Worker complaints
- Worker problems
- Wastes

These are your hot spots. Fix the most worrisome ones first. Ideally, systems should be enclosed, fugitive emissions eliminated, and waste generation minimized. Some changes are very inexpensive, such as the use of double valves and seals. Enclosing systems, minimizing emissions, eliminating fugitive emissions, and reducing generation and handling of waste would reduce most sources of potential human exposure.

Companies, though, have not always perceived these engineering controls to be as cost-effective as they are. As competitiveness in world markets continues its decline, this mentality is likely to get worse. The concept of a smooth-running, efficient facility with few breakdowns, few maintenance problems, and happy, productive workers is, remarkably, not "the American way."

III. COST-EFFECTIVE SOLUTIONS

Reduce Emissions and Wastes by Engineering Changes

An investment in engineering controls has enormous returns; there are few exposures, reducing the need for industrial hygiene monitoring. A reduction in the amount of hazardous wastes generated pays for itself in no time as the cost of hazardous waste disposal escalates. These kinds of engineering controls would always be favored as solutions.

However, without the ability to implement engineering changes, the soul of a program must prevent worker exposures by affording the following:

- Excellent ventilation
- Respiratory protection
- Skin protection
- Worker education
- Worker monitoring

Ensure Good Ventilation

As many occupational exposures occur by inhalation, the necessity of good ventilation is obvious. Emissions should be exhausted, remembering that the flow of air must be away from a worker's breathing zone. The air intake and outflow for a particular work area must be used in the way it was designed. Remodeling or partitioning an area may create pockets of stagnant air. Also, the human element must be taken into account—people open and close windows, open and shut doors, and plug up drafts or ducts with cardboard. Good ventilation must be ensured with *all* activities and in *all* seasons.

If processes are potentially hazardous and/or there are frequent upsets or ventilation breakdowns, a regrettable atmosphere can be created in more ways than one. It makes perfect sense to have good ventilation; if there is poor ventilation and something goes wrong, the lack of proper ventilation may be interpreted as negligence. As noted in Chapter 13, one out of every four persons develops cancer. Given today's litigious climate, if workers from a poorly ventilated workplace develop cancer, woe to the employer. Worker complaints about air quality must be taken seriously. A work force that feels undeserving of good quality air is not likely to be productive.

By and large, good ventilation is not costly and should be considered fundamental. Concerns regarding air quality encompass outdoor and indoor environments, factories, and office buildings.

Avoid Odors and Tight Buildings

Odors in the workplace upset people. Employers are urged to make every effort to avoid odors in a work area—even pleasant ones. Odors remind people that something is in the air and may set off a wave of hysteria that gets out of control. Entire buildings have been evacuated because of odors, some for months at a time.

If possible, chemical fumigation and extermination of pests should be avoided in areas where people work. If the benefit of fumigation outweighs the risk of alienating a work force, it is recommended that fumigation be conducted under the following conditions: all personnel on all shifts (including security and janitorial workers, employees and/or subcontractors) should be notified in advance of planned extermination or fumigation. It is recommended that fumigation be conducted by professionals, ideally when employees are not scheduled (e.g., Christmas shutdown) to be present. Afterward, the work area should be aired out until odor-free, before anyone returns. Carpeted offices in centrally ventilated buildings where the windows cannot be opened should not be treated: for example;

A large, carpeted office area was located in a centrally-ventilated, steel-and-glass building. Two hundred persons were at computers in carpeted cubicles. Extermination for "carpet fleas" was conducted Friday evening. On Monday morning, approximately 10 percent of the work force complained of nausea and/or headache, many of whom left the premises. Seven persons visited a "detoxification center" that, amazingly, was recommended by the local medical society. Sauna "treatments" in excess of $3,000 apiece were recommended as "detoxification," following which, workers were "banned from the building" as permanently disabled. All filed claims for permanent disability.

Elsewhere, municipal offices were sprayed, including carpets and furniture; it was not clear whether room dividers were sprayed. Janitors, who cleaned during the weekend, noted a nauseating smell. On Monday morning, many city workers were nauseated and by Tuesday morning, all had left the work area. It was necessary to relocate the entire municipal offices, even though in this building the windows opened; the unpleasant stench of the pesticide remained on the carpeting. A flurry of claims ensued, many with allegations of permanent injury, with at least two requests for disability retirements.

In these examples, the first company eventually moved out. The second moved out, had carpeting and all furnishings washed, moved in, moved out again, and had carpeting and furnishings replaced prior to moving back into the facilities — and in this building, the windows opened.

"Tight buildings" appear to be another source of distress for workers and employees alike. Since the 1970s, energy-efficient buildings with centralized services have not allowed workers to control their air flow, temperature, or purity. People resent not being in control of their own air supply, especially if it is unsatisfactory. Where there are large numbers of persons and inadequate ventilation, the air quickly becomes stale. The design of energy-efficient buildings generally does not take into account impediments to air flow posed by subdivisions such as cubicles, partitions, and shared computer stations. Room dividers, such as these promote localized pockets of poor airflow and pockets of stagnant air. Moreover, odors such as cigarette smoke, pesticides, carpet glue and the like, persist.

When furnishing or renovating offices, the area should be well-ventilated and aired out before people move in. Otherwise, odors or poor ventilation may trigger complaints of headache, fatigue, sluggishness, and sometimes eye irritation. Invariably, symptoms are cured by increased air flow and fresh air. However, the occasional employee may wish to be "detoxified" and claim temporary or permanent disablement.

In buildings such as these, highly emotional and vehement disputes may arise between smoking and non-smoking employees. Non-smokers radically oppose the odor of cigarette smoke, even if it is in the hall or the lunchroom. Here, too, employers would be well-advised to either declare a smoke-free workplace, particularly if general ventilation is less than optimal, or effectively segregate smokers from non-smokers.

A woman who herself was a recent ex-smoker was administrative assistant to the chairman of the board. Each had their own office. The chairman, a smoker who was continually trying to stop, would smoke off and on at times in his own office. The woman filed a claim based on the fact that her company had failed to provide her with a smoke-free environment, according to city law. She alleged 25 different complaints including headaches, watery eyes, lung damage, memory loss, skin rash, menopause, hemorrhoids, varicose veins and dandruff. She received a cash settlement of $25,000 (and that's no smoke).

As a result of experiences such as these, we recommend that buildings be kept well-ventilated, odors avoided, and should odors occur, employers take employee complaints seriously. It seems better to overreact than to discount an odor complaint.

Provide Protective Respiratory Equipment

Besides good, clean air, a company using chemicals *must* provide personal protective equipment appropriate to the worker's job. Fresh air-supplied respirators are relatively inexpensive, and if the job can be done safely with the use of air lines, they should be used. The appropriate use of fresh air, which enables workers to enjoy a clean atmosphere, is being widely used by even the most macho of workers and surely minimizes the potential for toxic exposure. Other items, such as cartridge-type respirators and dust masks, should be used as appropriate, but must be used as part of a training and monitoring program; otherwise they are not likely to be used correctly. The author would insist, at a minimum, on the use of dust masks whenever airborne fibers, soot, dust, or sawdust are present.

Provide Skin Protection

While industrial hygiene monitoring concentrates on workplace airborne threshold limit values, the skin as a means of exposure has not received sufficient attention. As toxic occurrences commonly involve workers who are splashed or drenched, emphasis on protecting the skin is needed. Face shields, goggles, coveralls, gloves, boots, and work shoes should be provided, laundered and maintained in good working order as appropriate, *by the company.* These items are cheap. How cheap? Disposable rubber gloves, for example, cost $.50 a pair. Used at a rate of one pair per shift, they would cost approximately $125.00 a year. Let us say a worker who handles epoxies without gloves develops a contact dermatitis, cannot return to work and requires vocational retraining. If he cannot return to his job, vocational retraining could theoretically cost $30,000, or the cost of 60,000 pairs of gloves. This would be enough to provide 24 workers with gloves for ten years.

When skin protection is used, it must be used sensibly. It is inappropriate to require somebody to wear tight-fitting gloves for an eight-hour shift: it is too hot and the hands sweat. For certain activities, using gloves may be appropriate; likewise, certain job activities can be altered so that a tool is used, such as a brush, to prevent direct contact between highly reactive materials, such as epoxies and glues and a person's skin.

Also, simple training and enforced work rules such as washing reactive substances from the skin immediately if an exposure occurs, can prevent the onset of a contact dermatitis, which may then be hard to treat.

Show Workers Respect

Ventilation improvements that provide excellent air quality and liberal use of protective equipment are relatively inexpensive investments that have big payoffs including:

- Protection of workers, and
- demonstration of the respect the company has for its work force

This type of consideration for a work force is beyond cost and is likely to be repaid.

IV. DESIGN NEW FACILITIES CORRECTLY

The concept of zero human exposure must be incorporated into the design of every new facility, as well as the concepts of easy maintenance, the use of intrinsically non-hazardous substances and minimal waste generation. Also, plants should be designed more realistically to anticipate human error, equipment failure, and Murphy's Law. Safety engineers and physician-toxicologists should participate in the design of facilities while the ideas are still on the drawing board.

V. LEARN AND COMMUNICATE

The most important area of communication for safety managers is to convince management that expenditures for safeguards are necessary. Generally, management will listen to and activate well-thought-out, well-reasoned programs. When management is refractory, a workplace toxics audit that incorporates work-related injury claims and/or utilizes outside experts as "friends of the family" can be an asset. Audits can save money if a toxics management program has runaway costs, difficulty in selecting priorities, loses its focus, or becomes unmanageable. An outside view may also assist in resolving labor-management disputes. The role of the people who handle the toxics — the workers — is discussed in the next chapter.

17

Practical Advice to Employers About Workers Who Handle Toxics

The worker who handles toxics is at a high risk of getting drenched in an accident thereby putting others at risk. The following guidelines reduce that risk.

I. PROVIDE THE RIGHT TOOLS FOR THE JOB

An illiterate immigrant worker came to be examined because of ugly, weeping lesions on his skin. As a factory janitor, he was responsible for removing wastes, which were not separated into hazardous and nonhazardous materials. He carted out about three times as much waste as there were dumpsters, and there was no mechanical garbage compactor. So this little fellow, trying really hard to do a good job, in his T-shirt, pants, and sneakers, jumped into the dumpsters and compacted the waste.

II. CHOOSE THE RIGHT WORKER FOR THE JOB

Several farm workers mixed and applied pesticides and fertilizers. They admitted that they could not read or write English or their native tongue, had no idea what they were mixing, or of what precautions should be taken. Instead of accurate information, they passed along incorrect pesticide superstitions so that, following a cold or flu, some believed they were incapacitated for life.

A major transportation company implemented the idea of random, mandatory drug-testing for employees. The first year that this policy was implemented, the number of work-related accidents fell from over 900 to about 60.

One of the most effective aspects of a toxics management program is choosing the right worker for a potentially hazardous job. The best, most capable worker, who has the ability commensurate with responsibility should be the one selected. Yet, in some companies, handling toxics or wastes is considered a dirty job and is given to uneducated or problematic workers. Selecting a substance abuser, or choosing an illiterate who cannot read a label or follow protocols makes no sense. It asks for, and finds, trouble.

III. PERFORM POTENTIALLY HAZARDOUS OPERATIONS WITH A FULL, RESTED CREW

Shift-work appears to be an important contributor to mishaps. Many serious incidents occur on swing or graveyard shifts on a Sunday night and involve individuals who have worked a full work week and are on overtime. Special, unusual, experimental, or potentially hazardous runs are best conducted by a well-rested work force of full complement and at times when an emergency can be well-managed.

IV. LEARN FROM WORKERS

Communication with workers is essential in order to minimize exposures. Workers should be involved in every aspect of toxics management, from planning to participation in audits and investigations.

Simplify Jobs

The man who does the job will make suggestions on how exposures can be reduced or how his job can be simplified safely. If the overall process works and the change is relatively minor, implement it. If the whole process needs reworking, involve the men who run the machinery.

Understand Procedures With "One-on-One" Training

When air measurements or circumstances indicate the need to reduce exposures by personal protective equipment, involve the worker in the decision. Safety staff need to understand each worker's job — not only "on paper" — they need to see how work is done at each work station and on each shift. Likewise, a worker should be observed doing various tasks so that activities that potentially expose him can be pointed out. At that time, a plan to prevent exposures can be developed. As a result of this one-on-one interaction, both the worker and the industrial hygienist will gain.

Make Protective Equipment a Joint Venture

Protective gear should suit the job and the worker. The type of gear should be a joint decision between the industrial hygienist and the worker and periodically be re-evaluated to see whether this decision was correct.

Understand People

If people work together as a team, they must be compatible. Some shifts, particularly swing or graveyard, can get sloppy and into mischief. Areas at high risk for big trouble are obvious. A plant that is continually beset with mechanical or personnel problems — ventilation breakdowns, emissions, maintenance problems, or worker problems are "hot spots" that need attention. Remember Murphy's Law — it was written for toxics.

Emphasize Neatness

An unkempt area will foster sloppy work practices and mistakes and result in exposures. A worker with leukemia, standing before a jury, saying, "Every day I was covered with the stuff, and nobody ever told me it was bad for me," will find very sympathetic ears. There should be procedures for changing into clean work clothes upon arrival, showering before leaving the area, and departing the plant in clean street clothes. A work area that looks neat and clean usually is.

V. ENSURE A PROPERLY TRAINED, INTERESTED WORK FORCE

A work force that is properly trained, educated, and interested will have fewer mishaps and be an asset. Toxics training should be part of safety and accident prevention programs; toxics education is as important as prevention of fire and explosion. While many companies have implemented training as required by law, these should not be *pro forma* run-throughs but should actually provide a worker with useful information. It is crucial to provide a forum by which one can ask questions until the information is understood. It is important for a worker to have a working knowledge of the potential toxicity of the odd things: mixtures, intermediates, or reactive by-products and, especially, wastes — which are often complex mixtures in strange combinations — as these entities are often involved in accidents and mishaps. A coffee klatch, plant "safety rounds" and plant toxics audits can all be used to stimulate discussion and questions. Actually, many people find toxicology and chemistry very interesting if presented in a lively and colorful way. Just learning proper procedures and hazards according to label information gives a person a very slanted and fearful impression. In contrast, programs that impart a balanced and realistic overview of our chemical background as well as what to expect will educate and reassure.

Remember that workers and unions tend to become very agitated when they learn that they are working with carcinogens. Labor/management disputes arise for many reasons, toxics management being a real or political issue in many of them. Educational courses geared for workers (and management) on carcinogens and the differences between effects in humans and those shown in animal experiments can go a long way in clarifying the toxic issues of the day and disarm a potentially explosive atmosphere. Good worker relations, superb communication, and mutual respect will go a long way toward minimizing exposures.

The worker needs information presented in a way he/she can absorb.

For multinational companies and those with workers from different cultures, procedures and information must be conveyed in terms of the worker's own culture. Special programs or procedure modifications may be required. Corporations with facilities worldwide may need to adjust to the human constraints at diverse locations.

VI. TEST AT INTERVALS

Periodically, workers should be observed on the job and tested in the classroom to ensure that they follow procedures, use protective equipment appropriately, eliminate exposures, and understand the substances they are handling. Both the work area and the worker should be clean.

VII. LEARN FROM MISTAKES

When incidents do happen, everyone should learn from them in a way that allows the person who made the mistake to retain his/her dignity, rather than be left feeling like a jerk. All workers (even relief workers) handling potentially toxic materials should participate in toxic audits and reviews of each other. They will learn enhanced safety from each other as well as how to communicate effectively. Since the toxics area is so confusing, no question should be regarded as dumb.

VIII. ENSURE EFFECTIVE COMMUNICATION BETWEEN PRODUCTION WORKERS, INDUSTRIAL HYGIENE, AND MEDICAL PEOPLE

The professional who is responsible for toxics education and training of the worker must be able to communicate effectively. As the safety professional makes his rounds, takes his measurements or attends worker safety meetings, he has to field questions from workers concerned about exposures and health—whether he wants to or not.

Therefore, he is often a link between the medical department and the worker. He needs to be secure, knowledgeable, and able to answer, "I'll check on that and get back to you," when he does not know.

The medical department also communicates closely with workers and their family doctors and acts as a link between the worker and the industrial hygiene staff concerning possible workplace exposures. The medical surveillance program is discussed in the next chapter.

18

How to Focus Medical Surveillance Programs on Toxics

KEY TOPICS

While industrial hygiene and medical surveillance programs can be extraordinarily expensive, they need not be. Focus is the key: the most cost effective programs use resources where they count—on potentially hazardous jobs. The ingenuity, of course, is to remove the hazard from the job, as discussed throughout the book. From our experience in toxics litigation, we have found that the more industrial hygiene, medical, incident and termination exams are focused toward toxics, the easier it is to reassure a person and to defend unworthy claims: "We measured and measured, and she was not exposed to anything in the air or in her body. Lorraine was fine when she left the company." It is a very sad commentary on our society— tragic, actually—but we encounter huge class-action claims alleging monumental damages, usually after a plant closed. Extensive investigation found no basis for many of these allegations. Realistically, a medical surveillance program has several objectives:

- Prevent health problems by constant monitoring
- Detect transient symptoms before permanent injury occurs
- Document efforts
- Document health status

Components of a medical surveillance program focused on toxics are as follows:

- Identify potentially hazardous jobs
- Conduct pre-hire medical and competency screening on job candidates
- Obtain samples for baseline biologic monitoring
- Record chemical contact history
- Record exposure history
- Special response to exposures at levels sufficient to produce an injury
- Overreact to questions, incidents, and injuries
- Interval surveillance
- Industrial hygiene monitoring
- Worker education and participation in medical surveillance
- Termination or retirement evaluations

I. IDENTIFY POTENTIALLY HAZARDOUS JOBS

Potentially hazardous jobs can be identified in several ways: on the basis of intrinsic hazards of substances used, if they are highly reactive, explosive, or readily absorbed through the skin; because there is potential exposure to large quantities of a low hazard substance, such as alcohol or mill tailings, or because the processes or substances used are new or unfamiliar. There are various ways that companies can keep track of what workers handle. A simple system will work the best; once identified, workers in these jobs become part of a medical/toxics surveillance program.

II. DIRECT PRE-HIRE MEDICAL AND COMPETENCY SCREENING TO HAZARDOUS SUBSTANCE HANDLING

An employee must be comfortable with the idea of working with potentially hazardous materials, fully literate, sensible, and secure about using protective equipment. If someone is afraid of toxics, nervous or intolerant of wearing respirators, or has any other reservation (including "chemophobia," fear of working with chemicals), this is not the person for the job. Past history should dwell upon prior jobs, occupations, and substances the person came in contact with, and whether or not they had a lot of accidents and mishaps. Someone who is always getting into trouble or is accident-prone may not be the person you want.

Workers who have substance abuse problems (or any conditions that impair judgment) should categorically not be allowed to handle toxic substances; that may harm them, expose other workers, or threaten a community. Accordingly, workers need pre-employment drug screening and should agree to be part of a random drug screening program.

Workers are needed who have good judgment and are able to act and respond appropriately in emergencies. Persons should not only be able to read but to understand and interpret labels of toxic materials as they are written in English. Many persons in today's work force, including articulate ones, can not read. An employee who cannot read labels, follow written instructions, or understand the content of protocols and warning labels, is not a good job candidate.

III. INCLUDE BIOLOGIC MONITORING

Archive Samples on Pre-hire Exams

New hires and longstanding employees enrolled in toxic surveillance programs are benefitted by biologic monitoring—that is, measurement of specific substances in blood or urine (Chapter 20). Archiving samples provides a baseline against which future exposures can be compared and controls overspending.

Analyze Samples When Exposures Occur

If an exposure occurs, comprehensive medical examinations and immediate biologic monitoring are encouraged. The exam should check for the known effects of the chemical in people as well as effects of its chemical presence. For example, given that nitroglycerin lowers blood pressure, one would check blood pressure in a worker exposed to nitroglycerin. Methemoglobin can be assayed in the blood.

I have found the main use of biological monitoring is the ability to reassure the worker with conviction based on objective data, demonstrating that the exposure was minimal and/or that no chemical was detected by a very sensitive test. Industrial hygiene monitoring including air sampling, skin wipes, and wipes of the work surfaces can be used in the same way following an incident. Also, these data can be used to estimate or quantify exposures.

IV. RECORD CHEMICAL CONTACT AND EXPOSURE HISTORIES

Routine chemical contacts as well as specific incidents should be part of all employees' medical files. In the event that any question arises later (e.g., thirty years later), a reliable history is available for chemicals worked with, as well as specific incidents, if any. If both biologic monitoring and prompt, competent medical care are available, documentation of whether an exposure occurred can actually be part of the record. Most of the time, this information will be reassuring.

V. PAY SPECIAL ATTENTION AND RESPONSE TO INCIDENTS

An exposure incident or questions and concerns of workers must be taken seriously by the company to the point of overreaction—but calmly, and with common sense. In the event of possible exposures, workers must be promptly evaluated and treated by knowledgeable, competent persons who can be a resource for their questions, provide appropriate treatment and/or management, and get them back to work, reassured, as swiftly as possible.

VI. RE-EXAMINE AND MONITOR PERIODICALLY

Subsequent examinations and monitoring should be conducted at intervals appropriate to the type of work and hazard of substances involved. Depending on potential job hazard, some companies evaluate at intervals ranging between 3 months and 24 months and again at job transfers.

VII. INDUSTRIAL
HYGIENE MONITORING

The medical and industrial hygiene departments should interact effectively with each other concerning any potential problem areas first noted by either department. They need to work together so that the medical department is alerted to the possibility of potential exposures in a particular work area; likewise, if workers in a particular location are voicing medical complaints, industrial hygienists can investigate and correct the problem.

VIII. WORKER EDUCATION
AND PARTICIPATION

A well-focused toxics surveillance program can be a resource to workers by teaching them in groups, as well as privately, and answering in private any concerns, they may not wish to express in public. Nurses are especially effective listeners. Workers should also be encouraged to identify any limitations in themselves that may interfere with safe work practices.

IX. TERMINATION/RETIREMENT
COMPREHENSIVE EVALUATIONS

This exam should be focused into the cumulative work, exposure and medical history, with objective testing and biologic monitoring.

It was reassuring for Lorraine to know that after working in chemical manufacture for thirty years, she left in good health.

X. INTERACTION AMONG SAFETY,
HYGIENE, MEDICAL DEPARTMENTS,
AND MANAGEMENT

In some companies, health-related components of a toxics management program are located in different departments. In the best programs, they work together in the interests of the worker's safety. In today's litigious climate, no company can afford departments that use toxics to acquire turf or as a base for political infighting. Industrial hygiene, health and safety people, and the medical department must maintain an open-door policy with each other as well as workers. If a worker complains of some tightness in his/her chest, and an industrial hygienist tells you, yes, there was some exposure to toluene diisocyanate, that brief communication saves a lot of time—and provides a prompt explanation to the worker. It also alerts the company to correct the circumstances that led to the exposure. Medical departments and community physicians also rely on industrial hygienists to provide information on potential exposures. The medical department, in turn, can alert the safety staff to possible injuries or complaints associated with a specific pilot plant, chemical mixture, waste process, or production area. These departments can also exchange information on health trends that are being reported from a particular industry.

XI. REFOCUS PROGRAMS AND ELIMINATE UNNECESSARY TESTS

Sometimes, management, medical departments, safety, and industrial hygiene professionals have dramatically different opinions over where their priorities should be and where money should be spent. At times they may wish to interface with other experts for advice or to resolve a dispute. It may be desirable to obtain an outside view of one's program with respect to its focus, omissions, costs, success, or interactions. Some companies do too much testing and too much monitoring, at great expense and to no one's benefit. An outside opinion from "a friend of the family" can help to pinpoint unnecessary expenditures. Workplace toxics audits utilizing outside experts can also be useful to help a program:

- Be more effective
- Order its priorities
- Assess its cost benefit
- Check its focus
- Reassure or resolve labor management disputes concerning toxics
- Detect trends in injured worker claims
- Identify omissions
- Identify new areas of concern
- Eliminate unnecessary testing
- Develop policies concerning carcinogens
- Develop policies concerning reproductive hazards and/or fetal protection
- Provide comparisons with other programs

In summary, a successful medical/toxics surveillance program keeps a worker on the job in good health, and provides maximal benefit at low cost.

Problems confronting injured workers are discussed in Part 5. The use of written and laboratory resources in toxicology are discussed in the following chapters.

Part 4

Resources in Toxicology

19

For Your Desk: Useful Sources of Information

I. MEDICAL TOXICOLOGY

There are many toxicology journals and books. I have found that the most useful ones for answering specific questions in toxicology concerning patients are the following:

- Proctor, Nick H., Hughes, James P., and Fischmann, Michael L. 1988. *Chemical Hazards of the Workplace.* Second edition. 1988. Philadelphia: J.B. Lippincott Company.

This small, useful book lists many common compounds in alphabetical order and gives human data in preference to animal data. Of special interest to physicians and those concerned about the toxicity of specific chemicals. Highly recommended. Medically oriented. If a physician were to have one book, this one would be it.

- *Physicians' Desk Reference (PDR).* Latest edition. Oradell, New Jersey: Medical Economics Company.

Remember to look up drug side-effects.

- American Conference of Governmental Industrial Hygienists (ACGIH), Inc. Current edition. *Documentation of the Threshold Limit Values and Biological Exposure Indexes.* Cincinnati: ACGIH.

This large book, in loose-leaf format, is frequently updated as new compounds are reviewed. It contains summaries supporting the ACGIH time-weighted average-threshold limit values (TWA TLVs). It also provides documentation of "Biologic Exposure Indices" (BEIs) for humans, as discussed in Chapter 20. Recommended.

- Daugaard, J. 1978. *Symptoms and Signs in Occupational Disease, A Practical Guide.* Copenhagen: Munksgaard International. Distributed by Year Book Medical Publishers, Inc.

This book is out of date and out of print, but it is a gem. Part One is organized by symptoms and signs, while Part Two, according to chemical or physical agents. Regrettably, a current edition is not available.

- Ellenhorn, Matthew J., and Barceloux, Donald G. 1988. *Medical Toxicology: Diagnosis and Treatment of Human Poisoning.* New York: Elsevier Science Publishing Company, Inc.

A large, good medical toxicology textbook. Some chapters, such as the one on Drugs of Abuse, are excellent.

II. MEDICINE

The Merck Manual is generally, a good overview of medical conditions with a nice, comprehensive section of toxicology.

- *The Merck Manual of Diagnosis and Therapy.* Latest edition. Berkow, Robert, and Fletcher, Andrew eds. Rahway, New Jersey: Merck Sharp and Dohme Research Laboratories.

A handy book with concise answers to puzzling symptoms is the following:

- DeGowin, Elmer L., and DeGowin, Richard L. Latest edition. *Bedside Diagnostic Examination.* New York: Macmillan Publishing Company, Inc.

III. INDUSTRIAL EMERGENCIES

This handy small book may be useful in helping a company to formulate a health and safety plan. Of interest to a facility that does not have a specific medical plan.

- LeFevre, Mark J. 1980. *First Aid Manual for Chemical Accidents for Use with Nonpharmaceutical Chemicals* English-language edition, ed. Ernest I. Becker, Ph.D. New York: Van Nostrand Reinhold.
- Kelley, Robert B. 1989. *Industrial Emergency Preparedness.* New York: Van Nostrand Reinhold.

This book walks the reader through the development of emergency plans for both basic and extraordinary events, such as earthquakes and floods, which require emergency preparedness. This book is intended for companies concerned about emergencies, as well as terrorism.

IV. LABORATORIES

- Pacific Toxicology Laboratories, Los Angeles, California, offers a wide variety of analyses of environmental "toxics" from human specimens. Continually updating. Willing to develop custom tests.
- BioTrace Laboratories, Salt Lake City, Utah, offers a broad selection of clinical tests, "toxics" and toxic metabolite analyses and drug abuse screens from human specimens such as blood, urine, fat and bone. In addition to biologic samples, this laboratory also processes specimens for industrial hygiene surveillance and analyzes environmental samples such as air, soil, vegetation, water, waste water, hazardous and solid wastes. Custom tests available. Routine analytes are listed in the Appendix.
- SmithKline Beecham Clinical Laboratories, various locations, offers comprehensive drug screening from blood and urine, therapeutic drug screens, broad spectrum of radioallergosorbent tests (RAST) allergy tests from blood.
- Allergy Testing Laboratories, Margate, Florida; Specialty Laboratories, Santa Monica, California. These laboratories offer RAST testing for isocyanates.

The next chapter is devoted to use of the laboratory in toxicology, with reference data.

V. MATERIAL SAFETY DATA SHEETS (MSDS)

Material Safety Data Sheets (MSDS) are prepared by the manufacturer and vary considerably in quality. Some are wrong and out of date. Some good MSDS samples follow, which are divided into nine sections:

- Section 1: Physical data
- Section 2: Ingredients
- Section 3: Fire/explosion hazard data
- Section 4: Health hazard data
- Section 5: Effects of overexposure (including

emergency and first aid procedures.
- Section 6: Reactivity data
- Section 7: Spill/leak procedures
- Section 8: Special protection information
- Section 9: Special precautions

As they vary in quality, the main usefulness of many MSDS is the identification of major ingredients and their concentrations (Section 2) and threshold limit values (Section 5).

The quality of Section 4, the health hazard data, varies considerably in that many MSDS list every finding in animal experiments (irrelevant to workers), and therefore, are not useful to a doctor who is evaluating a specific complaint. Another drawback of MSDS is that some do not list all ingredients because they are "inert" or "proprietary secrets." While an "inert" ingredient may not have activity for the specific use of a product (e.g., glue or solvent in a fungicide formulation), it may be "active" in causing a health effect, e.g., photosensitive dermatitis or defatting of skin.

What you should look for on the MSDS is the following:

- Do these sheets really represent what your patient used?
- What type of a product is it—gas, liquid or solid?
- Is it in a form that is likely to be inhaled or absorbed through the skin?
- Where is the material kept in the workplace in relation to the patient's job (e.g., in a warehouse in the next building from your patient)?
- How is it used? Is it used in a reaction with other substances?

Pay special attention to the brand name as well as the chemical name. Brand names can sound very catchy. Totally different brands, or even different trade names by the same manufacturer, can contain essentially the same ingredients. Conversely, chemical names may sound very different, but their structures may be similar (see Figure 19–1).

Good examples of MSDS for a dust, a solvent, and a human carcinogen can be found at the end of this chapter (see Fig 19–2). These samples were selected for apparent completeness, as well as clarity.

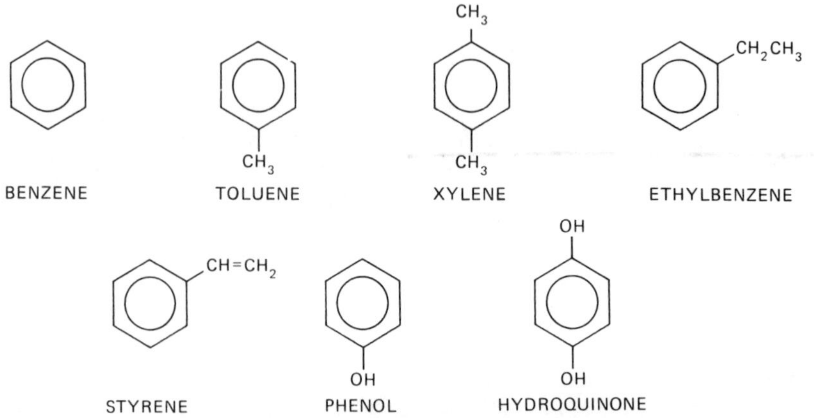

FIGURE 19-1. Chemicals of different structure with different sounding names

Material Safety Data Sheet

PPG INDUSTRIES, INC.	ONE PPG PLACE	PITTSBURGH, PA 15272

* * HI-SIL(R) 233

MSDS NUMBER:	0233	
DATE:	08/30/89	
EDITION:	008	
TRADE NAME:	HI-SIL(R) 233	
CHEMICAL NAME/SYNONYMS:	HYDRATED AMORPHOUS SILICA, SYNTHETIC PRECIPITATED SILICAS	
CHEMICAL FAMILY:	NONMETALLIC OXIDE	
FORMULA:	SI02	CAS NUMBER: 112926 00 8
U.S. DOT SHIPPING NAME:	NOT REGULATED	
U.S. DOT HAZARD CLASS:	NOT REGULATED	
SUBSIDIARY RISK:	N/A	
I.D. NUMBER:	N/A	
REPORTABLE QUANTITY:	N/A	
HMIS RATING:	0-0-0-E	

SECTION 1 * PHYSICAL DATA

BOILING POINT @ 760 MM HG:	N/A
VAPOR DENSITY (AIR = 1):	N/A
SPECIFIC GRAVITY (H20 = 1):	N/A
pH OF SOLUTIONS:	6.5–7.3 (5% SUSPENSION)
FREEZING/MELTING POINT:	N/A
SOLUBILITY (WEIGHT % IN WATER):	ESSENTIALLY INSOLUBLE
BULK DENSITY:	VARIABLE
VOLUME % VOLATILE:	N/A
VAPOR PRESSURE:	NONE
EVAPORATION RATE:	N/A
HEART OF SOLUTION:	N/A
APPEARANCE AND ODOR:	
WHITE, ODORLESS POWDER	

SECTION 2 * INGREDIENTS

MATERIAL	PERCENT
SI02 HYDRATE	87% (MIN)

CONTAINS NO DETECTABLE CRYSTALLITE SILICA—
(DETECTION LIMIT <0.01% BY WEIGHT)

SECTION 3 * FIRE/EXPLOSION HAZARD DATA

FLASH POINT (METHOD USED):
  NONE
FLAMMABLE LIMITS IN AIR (% BY VOLUME)
  LEL: N/A
  UEL: N/A
EXTINGUISHING MEDIA:
  N/A
SPECIAL FIRE FIGHTING PROCEDURES:
  N/A
UNUSUAL FIRE AND EXPLOSION HAZARDS:
  NONE

* * * 24-HOUR EMERGENCY ASSISTANCE: (304) 843-1300 * * *

FORM 6372 REV. 3/88

FIGURE 19-2. Material Safety Data Sheets (MSDS)

Material Safety Data Sheet

PPG INDUSTRIES, INC.	ONE PPG PLACE	PITTSBURGH, PA 15272

* * HI-SIL(R) 233   08/30/89   PAGE 2

SECTION 4 * HEALTH HAZARD DATA

TOXICITY DATA:
    LO50 INHALATION: SEE SECTION 5
    LD50 DERMAL: SEE SECTION 5
    SKIN/EYE IRRITATION: SEE SECTION 5
    LD50 INGESTION: ESTIMATED > 5 G/KG
    FISH,LD50 (LETHAL CONCENTRATION):   UNKNOWN

CLASSIFICATION: (POISON, IRRITANT, ETC.)
    INHALATION: NUISANCE DUST
    SKIN: SEE SECTION 5
    SKIN/EYE: MILDLY IRRITATING
    INGESTION: NOT SIGNIFICANTLY TOXIC
    AQUATIC: UNKNOWN

SECTION 5 * EFFECTS OF OVEREXPOSURE

THIS SECTION COVERS EFFECTS OF OVEREXPOSURE FOR INHALATION, EYE/SKIN CONTACT, INGESTION
AND OTHER TYPES OF OVEREXPOSURE INFORMATION IN THE ORDER OF THE MOST HAZARDOUS AND
THE MOST LIKELY ROUTE OF OVEREXPOSURE.

IS CHEMICAL LISTED AS A CARCINOGEN OR POTENTIAL CARCINOGEN?
NTP - NO   IARC - NO   OSHA - NO

MEDICAL CONDITIONS GENERALLY AGGRAVATED BY EXPOSURE:
    NONE KNOWN

PERMISSIBLE EXPOSURE LIMITS:
    OSHA: 6 MG/CU.M. (TOTAL DUST), 8-HOUR TWA (TIME WEIGHTED AVERAGE);
    29 CFR 1910.1000 (REV. 3/1/89).
    PPG INTERNAL PERMISSIBLE EXPOSURE LIMIT (IPEL):
    SYNTHETIC PRECIPITATED SILICAS: 5 MG/CU.M. (RESPIRABLE DUST), 8-HOUR TWA.

    ACUTE:
    EXCESSIVE CONTACT WITH POWDER CAN CAUSE DRYING OF MUCOUS MEMBRANES OF NOSE, EYES,
    AND THROAT DUE TO ABSORPTION OF MOISTURE AND OILS. THIS MATERIAL CAN ALSO CAUSE
    NASAL IRRITATION AND NOSEBLEEDS. EYE CONTACT WITH POWDER CAN RESULT IN MILD IRRITATION.

    CHRONIC:
    AN EPIDEMIOLOGICAL STUDY WAS CONDUCTED WHICH INCLUDED 165 PRECIPITATED SILICA WORKERS
    WHO HAD BEEN EXPOSED AN AVERAGE TIME SPAN OF 8.6 YEARS. OF THESE 165 WORKERS, 44 HAD
    BEEN EXPOSED FOR AN AVERAGE OF 18 YEARS. NO ADVERSE EFFECTS WERE NOTED IN COMPLETE
    MEDICAL EXAMINATIONS (INCLUDING CHEST ROENTGENOGRAMS) OF THESE WORKERS. PULMONARY
    FUNCTION DECREMENTS WERE CORRELATED ONLY WITH SMOKING AND AGE BUT NOT WITH THE
    DEGREE OR DURATION OF DUST EXPOSURES.

    LABORATORY STUDIES HAVE ALSO BEEN CONDUCTED IN SMALL ANIMALS VIA INHALATION TO
    LEVELS OF PRECIPITATED SILICA DUST OF UP TO 126 MG/CU.M. FOR PERIODS FROM SIX MONTHS TO
    TWO YEARS. ALTHOUGH PRECIPITATED SILICA WAS TEMPORARILY DEPOSITED IN THE ANIMALS'
    LUNGS, MOST OF THE DEPOSITED MATERIAL WAS CLEARED SOON AFTER THE DUST EXPOSURE ENDED.

    THE RESULTS OF ALL STUDIES PERFORMED BY, OR KNOWN TO, PPG INDICATE A VERY LOW ORDER
    OF PULMONARY ACTIVITY FOR SYNTHETIC PRECIPITATED SILICAS.

PPG RECOMMENDS THAT PERSONS WITH BREATHING PROBLEMS OR LUNG DISEASE SHOULD NOT WORK
    IN DUSTY AREAS UNLESS A PHYSICIAN APPROVES AND CERTIFIES THEIR FITNESS TO WEAR
    RESPIRATORY PROTECTION.

* * * 24-HOUR EMERGENCY ASSISTANCE: (304) 843-1300 * * *

FORM 6372 Rev. 3/88

Material Safety Data Sheet

PPG INDUSTRIES, INC.	ONE PPG PLACE	PITTSBURGH, PA 15272
* * HI-SIL(R) 233		08/30/89 PAGE 3

* EMERGENCY AND FIRST AID PROCEDURES

INHALATION:
   IF SYMPTOMS OF DISCOMFORT OR IRRITATION OCCUR DUE TO THIS PRODUCT, REMOVE
   AFFECTED PERSON TO FRESH AIR. IF IRRITATION OR DISCOMFORT PERSISTS, CONSULT A PHYSICIAN.
EYE OR SKIN CONTACT;
   IF POWDER ENTERS EYES, FLUSH WITH PLENTY OF WATER. IF IRRITATION OR DISCOMFORT OCCURS,
   CONSULT A PHYSICIAN.
INGESTION:
   NOT A LIKELY ROUTE OF EXPOSURE.
NOTES TO PHYSICIAN (INCLUDING ANTIDOTES):
   TREAT SYMPTOMATICALLY

SECTION 6 * REACTIVITY DATA

STABILITY: STABLE
   CONDITIONS TO AVOID: HIGH TEMPERATURES (>800 C) TREATMENT (CALCINING).
HAZARDOUS POLYMERIZATION: WILL NOT OCCUR
   CONDITIONS TO AVOID: NONE
INCOMPATIBILITY (MATERIALS TO AVOID)
   AVOID ALTERATION OF PRODUCT PROPERTIES BEFORE USE. CALCINING, WHICH MAY RESULT IN
   CRYSTALLINE FORMATION OR MIXING WITH ADDITIVES MAY ALTER TOXICOLOGICAL PROPERTIES.
HAZARDOUS DECOMPOSITION PRODUCTS:
   NONE.

SECTION 7 * SPILL OR LEAK PROCEDURES

STEPS TO BE TAKEN IF MATERIAL IS SPILLED OR RELEASED:
   VACUUM SPILLED MATERIAL AND PLACE IN CLOSED PLASTIC BAGS FOR DISPOSAL. SEE SECTION 8
   FOR PERSONAL PROTECTION INFORMATION.
WASTE DISPOSAL METHOD:
   WASTE FROM THIS PRODUCT MAY BE DISPOSED OF IN A SANITARY LANDFILL IF STATE AND LOCAL
   REGULATIONS PERMIT. CARE SHOULD BE EXERCISED TO AVOID CREATION OF DUST DURING DISPOSAL
   OPERATIONS.

SECTION 8 * SPECIAL PROTECTION INFORMATION

RESPIRATORY PROTECTION:
   USE NIOSH/MSHA APPROVED DUST FILTER RESPIRATOR FOR EXPOSURE ABOVE PERMISSIBLE EXPOSURE
   LIMITS. THE RESPIRATORY USE LIMITATIONS MADE BY NIOSH/MSHA OR THE MANUFACTURER MUST
   BE OBSERVED. RESPIRATORY PROTECTION PROGRAMS MUST BE IN ACCORDANCE WITH 29 CFR 1910.134.
VENTILATION(TYPE):
   GENERAL OR LOCAL EXHAUST SUFFICIENT TO MAINTAIN EMPLOYEE EXPOSURE BELOW PERMISSIBLE
   EXPOSURE LIMITS.
EYE PROTECTION:
   IF EYE EXPOSURE TO POWDER LIKELY, USE TIGHT FITTING PROTECTIVE GOGGLES.
GLOVES:
   CLOTH, LEATHER, OR RUBBER
OTHER PROTECTIVE EQUIPMENT:
   BOOTS, APRONS, OR CHEMICAL SUITS SHOULD BE USED WHEN NECESSARY TO PREVENT SKIN
   CONTACT. PERSONAL PROTECTIVE CLOTHING AND USE OF EQUIPMENT MUST BE IN ACCORDANCE
   WITH 29 CFR 1910.132 AND 29 CFR 1910.133.

* * * 24-HOUR EMERGENCY ASSISTANCE: (304) 843-1300 * * *

FORM 6372 Rev. 3/88

Material Safety Data Sheet

PPG INDUSTRIES, INC.	ONE PPG PLACE	PITTSBURGH, PA 15272

* * HI-SIL(R) 233   08/30/89   PAGE 5

SECTION 9 * SPECIAL PRECAUTIONS

PRECAUTIONS TO BE TAKEN DURING HANDLING AND STORING:
* * WEAR RESPIRATORY PROTECTION WHEN DUST EXPOSURE IS ABOVE PERMISSIBLE EXPOSURE LIMITS. RESPIRATORY PROTECTION MUST BE NIOSH/MSHA APPROVED DUST FILTER RESPIRATOR.
* * STORE IN DRY AREA.
OTHER PRECAUTIONS:
* * AVOID PROLONGED OR REPEATED INHALATION OF DUST. MAY IRRITATE THE RESPIRATORY TRACT.
* * AVOID CONTACT WITH EYES. MAY CAUSE IRRITATION AND PAIN.
* * AVOID PROLONGED, REPEATED, OR EXCESSIVE CONTACT WITH SKIN. MAY CAUSE IRRITATION AND DISCOMFORT.
* * USE WITH ADEQUATE VENTILATION. VENTILATION MUST BE SUFFICIENT TO LIMIT EMPLOYEE EXPOSURE TO THIS PRODUCT BELOW PERMISSIBLE EXPOSURE LIMITS. OBSERVANCE OF LOWER LIMITS (SECTION 5) IS ADVISABLE.
* * WASH THOROUGHLY EVERYDAY AFTER WORK.
* * DO NOT EAT, DRINK OR SMOKE IN WORK AREA.
COMMENTS:
TSCA - SYNTHETIC AMORPHOUS SILICA IS ON THE TSCA INVENTORY UNDER CAS NO. 112926-00-8.
SARA TITLE III - A) 311/312 CATEGORIES - ACUTE, B) NOT LISTED IN SECTION 313, C) NOT LISTED AS AN "EXTREMELY HAZARDOUS SUBSTANCE" IN SECTION 302.

R. KENNETH LEE
MANAGER, PRODUCT SAFETY

---

* * MONOCHLOROBENZENE

THIS MSDS/PSIS HAS BEEN REVIEWED FOR PROPER INFORMATION AS DEFINED BY OSHA HAZARD COMM. STD. 29 CFR 1910.1200; AND THE APPROPRIATE INFO.

MSDS NUMBER	0029	TRANSFERRED TO DATA SYSTEM.
DATE:	01/03/89	REVIEWED BY : _____ DATE: 2/21/90
EDITION:	006	
TRADE NAME:	MONOCHLOROBENZENE	
CHEMICAL NAME/SYNONYMS:	MCB, CHLOROBENZENE	
CHEMICAL FAMILY:	AROMATIC CHLOROHYDROCARBONS	
FORMULA:	C6H5CL	CAS NUMBER: 000108 90 7
U.S. DOT SHIPPING NAME:	MONOCHLOROBENZENE	
U.S. DOT HAZARD CLASS:	FLAMMABLE LIQUID	
SUBSIDIARY RISK:	N/A	
I.D. NUMBER:	UN1134	
REPORTABLE QUANTITY:	100 LBS/45.4 KGS	

SECTION 1 * PHYSICAL DATA

BOILING POINT @ 760 MM HG:	131.6 C (268.9 F)
VAPOR DENSITY (AIR = 1):	3.88
SPECIFIC GRAVITY (H20 = 1):	1.1071 @ 20 C
PH OF SOLUTIONS:	NEUTRAL
FREEZING/MELTING POINT:	−45.2 C (−49.4 F)
SOLUBILITY (WEIGHT % IN WATER):	.0488 @ 30 C
BULK DENSITY:	9.18 LBS/GAL @ 25 C
VOLUME % VOLATILE:	100

---

* * * 24-HOUR EMERGENCY ASSISTANCE: (304) 843-1300 * * *

FORM 6372 Rev. 3/88

Material Safety Data Sheet

PPG INDUSTRIES, INC.	ONE PPG PLACE	PITTSBURGH, PA 15272
* MONOCHLOROBENZENE		01/03/89 PAGE 2

VAPOR PRESSURE: 11.8 MM HG @ 20 C
EVAPORATION RATE: N/A
HEAT OF SOLUTION: NEGLIGIBLE
APPEARANCE AND ODOR:
   COLORLESS LIQUID WITH MILD MOTHBALL-LIKE ODOR

SECTION 2 * INGREDIENTS

MATERIAL	PERCENT
MONOCHLOROBENZENE	99.9

SECTION 3 * FIRE/EXPLOSION HAZARD DATA

FLASH POINT (METHOD USED):
   28 C (82.4 F) (TAG CLOSED CUP)
FLAMMABLE LIMITS IN AIR (% BY VOLUME)
   LEL: 1.3%
   UEL: 7.1%
EXTINGUISHING MEDIA:
   CARBON DIOXIDE, DRY CHEMICALS
SPECIAL FIRE FIGHTING PROCEDURES:
   FIRE FIGHTERS MUST WEAR NIOSH/MSHA-APPROVED, SELF-CONTAINED PRESSURE-DEMAND BREATHING APPARATUS.
UNUSUAL FIRE AND EXPLOSION HAZARDS:
   BURNS TO FORM TOXIC AND CORROSIVE HYDROGEN CHLORIDE GAS AND POSSIBLE TRACES OF PHOSGENE.

SECTION 4 * HEALTH HAZARD DATA

TOXICITY DATA:
   LC50 INHALATION: (RAT) 22,000 PPM
   LD50 DERMAL: SEE SECTION 5
   SKIN/EYE IRRITATION: SEE SECTION 5
   LD50 INGESTION: (RAT) 2910 MG/KG
   FISH,LC50 (LETHAL CONCENTRATION): TLM96: 100–1 PPM

CLASSIFICATION: (POISON, IRRITANT, ETC.)
   INHALATION: SLIGHTLY TOXIC
   SKIN: SEE SECTION 5
   SKIN/EYE: SEE SECTION 5
   INGESTION: SLIGHTLY TOXIC
   AQUATIC: TOXIC

SECTION 5 * EFFECTS OF OVEREXPOSURE

THIS SECTION COVERS EFFECTS OF OVEREXPOSURE FOR INHALATION, EYE/SKIN CONTACT, INGESTION AND OTHER TYPES OF OVEREXPOSURE INFORMATION IN THE ORDER OF THE MOST HAZARDOUS AND THE MOST LIKELY ROUTE OF OVEREXPOSURE.

IS CHEMICAL LISTED AS A CARCINOGEN OR POTENTIAL CARCINOGEN?
NTP - NO   IARC - NO   OSHA - NO

MEDICAL CONDITIONS GENERALLY AGGRAVATED BY EXPOSURE:
   NONE KNOWN

* * * 24-HOUR EMERGENCY ASSISTANCE: (304) 843-1300 * * *

FORM 6372 Rev. 3/88

Material Safety Data Sheet

PPG INDUSTRIES, INC.	ONE PPG PLACE	PITTSBURGH, PA 15272

* MONOCHLOROBENZENE   01/03/89   PAGE 3

PERMISSIBLE EXPOSURE LIMITS:
OSHA: 75 PPM (350 MG/CU.M.), 8-HOUR TWA (TIME-WEIGHTED AVERAGE); 29 CFR 1910.1000.
PPG INTERNAL PERMISSIBLE EXPOSURE LIMIT (IPEL): 75 PPM, 8-HOUR TWA (TIME-WEIGHTED AVERAGE);
150 PPM, STEL (SHORT-TERM EXPOSURE LIMIT).

ACUTE:
INHALATION: MONOCHLOROBENZENE (MCB) IS A CENTRAL NERVOUS SYSTEM DEPRESSANT.
INHALATION AT CONCENTRATIONS IN EXCESS OF THE OSHA PERMISSIBLE EXPOSURE LIMIT CAN
CAUSE HEADACHE, DIZZINESS, EYE, NOSE AND THROAT IRRITATION, NAUSEA, FEELING OF
DRUNKENNESS, UNCONSCIOUSNESS AND EVEN DEATH IN CONFINED OR POORLY VENTILATED AREAS.
THE AVAILABLE SCIENTIFIC LITERATURE INDICATES THAT SEVERE INHALATION OVEREXPOSURE MAY
RESULT IN LIVER AND KIDNEY INJURY.

EYE/SKIN: ANIMAL STUDIES AND HUMAN EXPERIENCE INDICATE THAT SHORT EXPOSURES CAN
RESULT IN MINOR EYE AND SKIN IRRITATION AND PAIN; HOWEVER, PROLONGED OR REPEATED
CONTACT MAY RESULT IN MILD SKIN BURNS.

INGESTION: ALTHOUGH SWALLOWING IS NOT A LIKELY ROUTE OF EXPOSURE IN INDUSTRIAL
APPLICATIONS, ACCIDENTAL INGESTION OF LARGE QUANTITIES OF MCB CAN CAUSE ILLNESS SIMILAR
TO THOSE DESCRIBED ABOVE FOR INHALATION.

CHRONIC:
VARIOUS LONG-TERM MCB STUDIES (INHALATION AND INGESTION) INVOLVING SEVERAL SPECIES OF
EXPERIMENTAL ANIMALS HAVE REPORTED CHANGES IN THE LIVER, KIDNEYS, AND BLOOD.

MONOCHLOROBENZENE WAS ADMINISTERED TO MICE AND RATS AT DAILY ORAL DOSAGE OF 60, 125,
250, 500 AND 750 MG/KG FOR 5 DAYS/WEEK IN A THREE-MONTH TOXICITY STUDY CONDUCTED BY
THE NATIONAL TOXICOLOGY PROGRAM. DOSES OF 250 MG/KG OR HIGHER CAUSED REDUCED BODY
WEIGHTS AND/OR DECREASED SURVIVAL. CHANGES IN CLINICAL CHEMISTRY, HEMATOLOGY, URINE
CHEMISTRY AND/OR ORGAN WEIGHTS WERE OBSERVED IN RATS. ORGAN WEIGHT CHANGES WERE
OBSERVED IN MICE AT DOSES OF 250 MG/KG OR HIGHER. DOSE-RELATED HEPATOCELLULAR NECROSIS
WAS OBSERVED IN BOTH RATS AND MICE. LYMPHOID OR MYELOID DEPLETION OF THE BONE
MARROW, THYMUS AND SPLEEN AND NEPHROTOXICITY WERE OBSERVED AT OR ABOVE DOSES OF
250 MG/KG IN MICE AND 500 MG/KG IN RATS.

IN A CHRONIC BIOASSAY CONDUCTED BY THE NATIONAL TOXICOLOGY PROGRAM, DAILY ORAL
DOSES (5 DAYS/WEEK FOR 103 WEEKS) OF 60 OR 120 MG/KG WERE ADMINISTERED TO MALE AND
FEMALE RATS AND FEMALE MICE AND TO MALE MICE AT DOSES OF 30 OR 60 MG/KG. DECREASED
SURVIVAL AND NEOPLASTIC NODULES OF THE LIVER WERE OBSERVED IN HIGH-DOSE MALE RATS
BUT NO INCREASE IN NODULES OR TUMOR INCIDENCE WAS REPORTED IN FEMALE RATS OR MALE
OR FEMALE MICE. NO LYMPHOID OR MYELOID DEPLETION OF BONE MARROW, THYMUS OR SPLEEN
WAS OBSERVED, IN CONTRAST TO THE 90-DAY STUDY RESULTS. NO NON-NEOPLASTIC LESIONS WERE
CLEARLY ATTRIBUTABLE TO MONOCHLOROBENZENE.

IN AN INHALATION (50, 150, AND 450 PPM) TWO-GENERATION REPRODUCTION STUDY IN RATS, IN THE
LOW-DOSE GROUP, NO ADVERSE EFFECTS WERE EVIDENT EXCEPT THAT THE RELATIVE LIVER TO
BODY WEIGHT RATIO WAS STATISTICALLY HIGHER THAN CONTROL. IN THE MID AND HIGH-DOSE
GROUPS, NO ADVERSE EFFECTS WERE EVIDENT IN BODY WEIGHT, FOOD CONSUMPTION, PHYSICAL
OBSERVATIONS, REPRODUCTIVE PERFORMANCE, OR FERTILITY, IN THE ADULT GENERATIONS. AT
SACRIFICE OF THE MID AND HIGH-DOSE ADULTS, RELATIVE LIVER WEIGHTS WERE HIGHER THAN
CONTROL. MICROSCOPIC EXAMINATION OF TISSUES FOR F0 AND F1 ADULTS REVEALED
HEPATOCELLULAR HYPERTROPHY AND DEGENERATIVE AND INFLAMMATORY RENAL LESIONS AND
AN INCREASE IN THE INCIDENCE OF TESTICULAR DEGENERATIVE CHANGES IN THE HIGH-DOSE MALES
AND IN THE F1 MID-DOSE MALES. NO ADVERSE EFFECTS WERE EVIDENT FROM THE EVALUATION OF
PUPS DELIVERED, NURSED, AND WEANED TO THESE SAME FEMALES.

* * * 24-HOUR EMERGENCY ASSISTANCE: (304) 843-1300 * * *

FORM 6372 Rev. 3/88

Material Safety Data Sheet

PPG INDUSTRIES, INC.	ONE PPG PLACE	PITTSBURGH, PA 15272

* MONOCHLOROBENZENE   01/03/89   PAGE 4

ALTHOUGH RESULTS OF CHRONIC ANIMAL STUDIES ARE INDICATIVE OF POTENTIAL ADVERSE EFFECTS IN HUMANS, THESE EFFECTS HAVE NOT BEEN OBSERVED IN HUMANS.

OTHER STUDIES HAVE SHOWN MONOCHLOROBENZENE TO BE WITHOUT TERATOGENIC (BIRTH DEFECTS) OR MUTAGENIC EFFECTS.

* EMERGENCY AND FIRST AID PROCEDURES

INHALATION:
REMOVE TO FRESH AIR. IF NOT BREATHING, GIVE ARTIFICIAL RESPIRATION, PREFERABLY MOUTH-TO-MOUTH. IF BREATHING IS DIFFICULT, GIVE OXYGEN. CALL A PHYSICIAN.
EYE OR SKIN CONTACT:
FLUSH EYES AND SKIN WITH PLENTY OF WATER (SOAP AND WATER FOR SKIN) FOR AT LEAST 15 MINUTES, WHILE REMOVING CONTAMINATED CLOTHING AND SHOES. IF IRRITATION OCCURS, CONSULT A PHYSICIAN.
INGESTION:
IF CONSCIOUS, DRINK LARGE QUANTITIES OF WATER. DO NOT INDUCE VOMITING. TAKE IMMEDIATELY TO A HOSPITAL OR PHYSICIAN. IF UNCONSCIOUS, OR IN CONVULSIONS, TAKE IMMEDIATELY TO A HOSPITAL. DO NOT ATTEMPT TO INDUCE VOMITING OR GIVE ANYTHING BY MOUTH TO AN UNCONSCIOUS PERSON.

NOTES TO PHYSICIAN (INCLUDING ANTIDOTES):
NEVER ADMINISTER ADRENALINE FOLLOWING MONOCHLOROBENZENE OVEREXPOSURE. INCREASED SENSITIVITY TO THE HEART TO ADRENALINE MAY BE CAUSED BY OVEREXPOSURE TO MONOCHLOROBENZENE.

SECTION 6 * REACTIVITY DATA

STABILITY: STABLE
CONDITIONS TO AVOID: OPEN FLAMES, HOT GLOWING SURFACES OR ELECTRIC ARCS.
HAZARDOUS POLYMERIZATION: WILL NOT OCCUR
CONDITIONS TO AVOID: NONE
INCOMPATIBILITY (MATERIALS TO AVOID):
NONE
HAZARDOUS DECOMPOSITION PRODUCTS:
HYDROGEN CHLORIDE AND POSSIBLE TRACES OF PHOSGENE

SECTION 7 * SPILL OR LEAK PROCEDURE

STEPS TO BE TAKEN IF MATERIAL IS SPILLED OR RELEASED:
IMMEDIATELY EVACUATE THE AREA, PROVIDE MAXIMUM VENTILATION AND REMOVE ALL SOURCES OF IGNITION. UNPROTECTED PERSONNEL SHOULD MOVE UPWIND OF SPILL. ONLY PERSONNEL EQUIPPED WITH NIOSH/MSHA-APPROVED, SELF-CONTAINED BREATHING APPARATUS OR FULL FACEPIECE AIRLINE RESPIRATORS WITH AUXILIARY SCBA'S OPERATED IN THE PRESSURE-DEMAND MODE AND EYE/SKIN PROTECTION SHOULD BE PERMITTED IN AREA. DIKE AREA TO CONTAIN SPILL. TAKE PRECAUTIONS AS NECESSARY TO PREVENT CONTAMINATION OF GROUND AND SURFACE WATERS. RECOVER SPILLED MATERIAL ON ADSORBENTS, SUCH AS SAWDUST OR VERMICULITE, AND SWEEP INTO CLOSED CONTAINERS FOR DISPOSAL. ONLY AFTER ALL VISIBLE TRACES HAVE BEEN REMOVED AND ABOVE CLEAN-UP STEPS PERFORMED, THOROUGHLY WET VACUUM THE AREA. DO NOT FLUSH TO SEWER. IF AREA OF SPILL IS POROUS, REMOVE AS MUCH CONTAMINATED EARTH AND GRAVEL, ETC., AS NECESSARY AND PLACE IN CLOSED CONTAINERS FOR DISPOSAL.

* * * 24-HOUR EMERGENCY ASSISTANCE: (304) 843-1300 * * *

FORM 6372 Rev. 3/88

Material Safety Data Sheet

PPG INDUSTRIES, INC.	ONE PPG PLACE	PITTSBURGH, PA 15272

* MONOCHLOROBENZENE   01/03/89   PAGE 5

WASTE DISPOSAL METHOD:
   CONTAMINATED SAWDUST, VERMICULITE, OR POROUS SURFACE AND/OR WET VACUUMED MATERIAL
   MUST BE DISPOSED OF IN AN APPROVED HAZARDOUS WASTE MANAGEMENT FACILITY. CARE MUST
   BE TAKEN WHEN USING OR DISPOSING OF CHEMICAL MATERIALS AND/OR THEIR CONTAINERS TO
   PREVENT ENVIRONMENTAL CONTAMINATION. IT IS YOUR DUTY TO DISPOSE OF THE CHEMICAL
   MATERIALS AND/OR THEIR CONTAINERS IN ACCORDANCE WITH THE CLEAN AIR ACT, THE CLEAN
   WATER ACT, THE RESOURCE CONSERVATION AND RECOVERY ACT, AS WELL AS ANY OTHER
   RELEVANT FEDERAL, STATE OR LOCAL LAWS/REGULATIONS REGARDING DISPOSAL.

RCRA—HAZARDOUS WASTE NUMBER U-037

SECTION 8 * SPECIAL PROTECTION INFORMATION

RESPIRATORY PROTECTION
   USE NIOSH/MSHA-APPROVED ORGANIC VAPOR CARTRIDGE OR CANISTER RESPIRATOR FOR ROUTINE
   WORK PURPOSES WHEN AIR CONCENTRATIONS EXCEED THE PERMISSIBLE EXPOSURE LIMITS. SEE
   SECTION 7 FOR RESPIRATORY PROTECTION DURING SPILLS OR LEAKS. THE RESPIRATOR USE
   LIMITATIONS MADE BY NIOSH/MSHA OR THE MANUFACTURER MUST BE OBSERVED. RESPIRATORY
   PROTECTION PROGRAMS MUST BE IN ACCORDANCE WITH 29 CFR 1910.134.
VENTILATION(TYPE):
   LOCAL EXHAUST VENTILATION SUFFICIENT TO MAINTAIN WORKPLACE CONCENTRATION BELOW
   PERMISSIBLE EXPOSURE LIMITS.
EYE PROTECTION:
   CHEMICAL SAFETY GOGGLES
GLOVES:
   VITON(R). LIMITED SERVICES ONLY: NITRILE BUTYL RUBBER.
OTHER PROTECTIVE EQUIPMENT:
   IMPERVIOUS BOOTS, APRONS, OR CHEMICAL SUITS SHOULD BE USED WHEN NECESSARY TO PREVENT
   SKIN CONTACT. PERSONAL PROTECTIVE CLOTHING AND USE OF EQUIPMENT MUST BE IN
   ACCORDANCE WITH 29 CFR 1910.132 AND 29 CFR 1910.133.

SECTION 9 * SPECIAL PRECAUTIONS

PRECAUTIONS TO BE TAKEN DURING HANDLING AND STORING:
   * CONTAINER AND SYSTEM MUST BE ELECTRICALLY GROUNDED BEFORE UNLOADING.
   * ADEQUATE VENTILATION MUST BE MAINTAINED IN STORAGE AREA TO REDUCE FIRE HAZARD IN
     THE EVENT OF A LEAK.
   * STORE ONLY IN CLOSED, PROPERLY LABELED CONTAINERS.
   * DO NOT USE IN CONFINED OR POORLY VENTILATED AREAS.
   * THIS MATERIAL OR ITS VAPORS WHEN IN CONTACT WITH FLAMES, HOT GLOWING SURFACES OR
     ELECTRIC ARCS, CAN DECOMPOSE OR BURN TO FORM HYDROGEN CHLORIDE GAS AND POSSIBLE
     TRACES OF PHOSGENE.
   * KEEP AWAY FROM HEAT, SPARKS, OR FLAMES.
   * AVOID CONTAMINATION OF WATER SUPPLIES! HANDLING, STORAGE AND USE PROCEDURES MUST
     BE CAREFULLY MONITORED TO AVOID SPILLS OR LEAKS. ANY SPILL OR LEAK HAS THE POTENTIAL
     TO CAUSE UNDERGROUND WATER CONTAMINATION WHICH MAY, IF SUFFICIENTLY SEVERE,
     RENDER A DRINKING WATER SOURCE UNFIT FOR HUMAN CONSUMPTION. CONTAMINATION THAT
     DOES OCCUR CANNOT BE EASILY CORRECTED.
OTHER PRECAUTIONS
   * DO NOT BREATHE VAPORS. HIGH VAPOR CONCENTRATION CAN CAUSE DIZZINESS,
     UNCONSCIOUSNESS, OR DEATH. LONG-TERM OVEREXPOSURE MAY CAUSE LIVER/KIDNEY INJURY
     AND OTHER ORGAN DAMAGE.
   * USE ONLY WITH VENTILATION SUFFICIENT TO KEEP EMPLOYEE EXPOSURES BELOW PERMISSIBLE
     EXPOSURE LIMITS. OBSERVANCE OF LOWER LIMITS (SEE SECTION 5) IS ADVISABLE.
   * AVOID CONTACT WITH EYES. WILL CAUSE IRRITATION AND PAIN.

* * * 24-HOUR EMERGENCY ASSISTANCE: (304) 843-1300 * * *

FORM 6372 Rev. 3/88

Material Safety Data Sheet

PPG INDUSTRIES, INC.	ONE PPG PLACE	PITTSBURGH, PA 15272
• MONOCHLOROBENZENE		01/03/89 PAGE 6

* AVOID PROLONGED OR REPEATED CONTACT WITH SKIN. MAY CAUSE IRRITATION OR DERMATITIS.
* DO NOT SWALLOW. SWALLOWING MAY CAUSE INJURY OR DEATH.
* DO NOT EAT, DRINK, OR SMOKE IN WORK AREAS.

COMMENTS:
TSCA - MONOCHLOROBENZENE IS ON THE TSCA INVENTORY UNDER CAS #108-90-7.

SARA TITLE III - A) 311/312 CATEGORIES - ACUTE, CHRONIC AND FLAMMABILITY, B) LISTED IN SECTION 313 UNDER CHLOROBENZENE, C) NOT LISTED AS AN "EXTREMELY HAZARDOUS SUBSTANCE" IN SECTION 302.

CERCLA - LISTED IN TABLE 302.4 OF 40 CFR PART 302 AS A HAZARDOUS SUBSTANCE WITH A REPORTABLE QUANTITY OF 100 POUNDS. RELEASES TO AIR, LAND OR WATER WHICH EXCEED THE RQ MUST BE REPORTED TO THE NATIONAL RESPONSE CENTER, 800-424-8802.

RCRA - WASTE CHLOROBENZENE AND CONTAMINATED SOILS/MATERIALS FROM SPILL CLEANUP AND U037 HAZARDOUS WASTE AS PER 40 CFR 261.33 AND MUST BE DISPOSED OF ACCORDINGLY UNDER RCRA. SEE 40 CFR 261.33(C) AND 261.7(B)(3) FOR CLEANING REQUIREMENTS FOR EMPTY CONTAINERS.

CANADIAN WHMIS - A) SENSITIZATION TO PRODUCT: NONE KNOWN, B) REPRODUCTIVE TOXICITY: NONE KNOWN, C) ODOR THRESHOLD: NOT KNOWN, D) PRODUCT USE: REACTANT.

R. KENNETH LEE
MANAGER, PRODUCT SAFETY

---

* * * VINYL CHLORIDE (VCM)

MSDS NUMBER	0129	
DATE:	08/20/88	
EDITION:	005	
TRADE NAME:	VINYL CHLORIDE (VCM)	
CHEMICAL NAME/SYNONYMS:	VINYL CHLORIDE MONOMER	
CHEMICAL FAMILY:	CHLORINATED HYDROCARBON	
FORMULA:	CH2 = CHCL	CAS NUMBER: 0075 01 4
U.S. DOT SHIPPING NAME:	VINYL CHLORIDE	
U.S. DOT HAZARD CLASS:	FLAMMABLE GAS	
SUBSIDIARY RISK:	N/A	
I.D. NUMBER:	UN1086	
REPORTABLE QUANTITY:	1 LB.	

SECTION 1 • PHYSICAL DATA

BOILING POINT @ 760 MM HG:	7.0 F
VAPOR DENSITY (AIR = 1):	2.15
SPECIFIC GRAVITY (H20 = 1):	0.9121 @ 20/4 C
PH OF SOLUTIONS:	APPROX. 7
FREEZING/MELTING POINT:	−245 F (−153.7 C)
SOLUBILITY (WEIGHT % IN WATER):	0.11% @ 25 C (77 F)
BULK DENSITY:	APPROX. 7.59 LBS/GAL
VOLUME % VOLATILE:	100%
VAPOR PRESSURE:	2580 MM HG @ 20 C
EVAPORATION RATE:	(ETHYL ETHER = 1): < 1
HEAT OF SOLUTION:	N/A
APPEARANCE AND ODOR:	

CLEAR, COLORLESS, LIQUIFIED GAS UNDER PRESSURE; MILD, SWEET ODOR

---

* * * 24-HOUR EMERGENCY ASSISTANCE: (304) 843-1300 * * *

FORM 6372 Rev. 3/88

Material Safety Data Sheet

PPG INDUSTRIES, INC.	ONE PPG PLACE	PITTSBURGH, PA 15272
* * VINYL CHLORIDE (VCM)		08/20/88 PAGE 2

SECTION 2 * INGREDIENTS

MATERIAL   PERCENT
VINYL CHLORIDE MONOMER   100

SECTION 3 * FIRE/EXPLOSION HAZARD DATA

FLASH POINT (METHOD USED):
  −108 F (CLEVELAND OPEN CUP) (−78 C)

FLAMMABLE LIMITS IN AIR (% BY VOLUME)
  LEL: 3.6%
  UEL: 33%
EXTINGUISHING MEDIA:
  CARBON DIOXIDE OR DRY CHEMICAL FOR SMALL FIRES.

SPECIAL FIRE FIGHTING PROCEDURES:
  FOR ANY VCM FIRE, IMMEDIATELY EVACUATE AREA. VINYL CHLORIDE GAS IS HEAVIER THAN AIR.
  ONLY FIREFIGHTERS EQUIPPED WITH PRESSURE DEMAND, SELF-CONTAINED BREATHING APPARATUS
  SHOULD BE ALLOWED IN AREA. USE WATER SPRAY TO COOL EQUIPMENT. (ALSO SEE COMMENTS AT
  END OF MSDS)
UNUSUAL FIRE AND EXPLOSION HAZARDS:
  BURNS TO FORM TOXIC AND CORROSIVE GASES, MOSTLY HYDROGEN CHLORIDE.

SECTION 4 * HEALTH HAZARD DATA

TOXICITY DATA:
  LC50 INHALATION: SEE SECTION 5
  LD50 DERMAL: SEE SECTION 5
  SKIN/EYE IRRITATION: SEE SECTION 5
  LD50 INGESTION: SEE SECTION 5
  FISH,LC50 (LETHAL CONCENTRATION):   UNKNOWN

CLASSIFICATION: (POISON, IRRITANT, ETC.)
  INHALATION: SEE SECTION 5
  SKIN   SEE SECTION 5
  SKIN/EYE   SEE SECTION 5
  INGESTION   SEE SECTION 5
  AQUATIC: UNKOWN

SECTION 5 * EFFECTS OF OVEREXPOSURE

THIS SECTION COVERS EFFECTS OF OVEREXPOSURE FOR INHALATION, EYE/SKIN CONTACT, INGESTION
AND OTHER TYPES OF OVEREXPOSURE INFORMATION IN THE ORDER OF THE MOST HAZARDOUS AND
THE MOST LIKELY ROUTE OF OVEREXPOSURE.

IS CHEMICAL LISTED AS A CARCINOGEN OR POTENTIAL CARCINOGEN?
NTP - YES   IARC - YES   OSHA - YES

MEDICAL CONDITIONS GENERALLY AGGRAVATED BY EXPOSURE:
  NONE KNOWN.

PERMISSIBLE EXPOSURE LIMITS:
  OSHA: 1 PPM, 8-HOUR TWA (TIME-WEIGHTED AVERAGE); 5 PPM AVERAGED OVER ANY 15 MINUTE
  PERIOD, STEL (SHORT-TERM EXPOSURE LIMIT); ACTION LEVEL - 0.5 PPM, 8-HOUR TWA; 29 CFR 1910.1017.

* * * 24-HOUR EMERGENCY ASSISTANCE: (304) 843-1300 * * *

FORM 6372 Rev. 3/88

Material Safety Data Sheet

PPG INDUSTRIES, INC.	ONE PPG PLACE	PITTSBURGH, PA 15272

* * VINYL CHLORIDE (VCM)   08/20/88 PAGE 3

ACUTE:

INHALATION: ACUTE INHALATION OVEREXPOSURE TO VINYL CHLORIDE CAN CAUSE LUNG IRRITATION, DIZZINESS, AND UNCONSCIOUSNESS FOLLOWING SHORT-TERM EXPOSURE.

SKIN/EYE: SKIN AND EYE CONTACT WITH THE LIQUID MONOMER CAN CAUSE FREEZING OF EXPOSED TISSUES BECAUSE OF LOW STORAGE TEMPERATURES REQUIRED FOR THIS PRODUCT.

CHRONIC:

VINYL CHLORIDE HAS BEEN DEMONSTRATED TO BE A HUMAN CARCINOGEN. OCCUPATIONAL EXPOSURES TO VINYL CHLORIDE MONOMER HAVE BEEN ASSOCIATED WITH A VARIETY OF HUMAN DISEASE CONDITIONS. AMONG THESE ARE LIVER LESIONS, INCLUDING FIBROSIS AND CANCER (ANGIOSARCOMA AND OTHER CANCERS OF THE LIVER AND BILIARY TRACT), SKIN CONDITIONS (SCLERODERMA), AND SHORTENING OF THE FINGERS (ACROOSTEOLYSIS). OTHER ABNORMALITIES LINKED WITH VINYL CHLORIDE MONOMER EXPOSURES INCLUDE SPLEEN, LUNG, BLOOD, AND NERVOUS SYSTEM ABNORMALITIES (INCLUDING CANCER OF THE BRAIN AND OTHER CNS PARTS). IN A RECENT DRAFT REPORT ENTITLED "EPIDEMIOLOGIC STUDY OF VINYL CHLORIDE WORKERS," THE STUDY FOUND A MORTALITY EXCESS IN EMPHYSEMA, HOWEVER, THE STUDY DID NOT FIND ANY EXCESS IN EITHER RESPIRATORY CANCER OR LYMPHATIC AND HEMATOPOIETIC CANCER.

* EMERGENCY AND FIRST AID PROCEDURES

INHALATION:
REMOVE TO FRESH AIR. IF NOT BREATHING, GIVE ARTIFICIAL RESPIRATION, PREFERABLY MOUTH-TO-MOUTH. IF BREATHING IS DIFFICULT, GIVE OXYGEN. CALL A PHYSICIAN.
EYE OR SKIN CONTACT:
EYE CONTACT - FLUSH IMMEDIATELY WITH WATER FOR AT LEAST 15 MINUTES. CALL A PHYSICIAN. SKIN CONTACT - WASH WITH WATER, THEN SOAP AND WATER WHILE REMOVING CONTAMINATED CLOTHING AND SHOES. VINYL CHLORIDE'S RAPID EVAPORATION RATE CAN PRODUCE FROSTBITE. FOR FROSTBITE, DO NOT RUB BUT WARM AFFECTED AREA BY PLACING IN LUKEWARM WATER. EXERCISE AFFECTED AREA AFTERWARDS. GIVE PATIENT WARM DRINK. IF IRRITATION PERSISTS, CONTACT A PHYSICIAN. THOROUGHLY CLEAN CONTAMINATED CLOTHING BEFORE REUSE OR DISCARD.
INGESTION:
NOT APPLICABLE
NOTES TO PHYSICIAN (INCLUDING ANTIDOTES):
IT SHOULD BE NOTED THAT ENZYMATIC LIVER FUNCTION TESTS ARE NOT ABNORMAL IN INDIVIDUALS WITH VINYL CHLORIDE-ASSOCIATED LIVER LESIONS. ALTERNATE DIAGNOSTIC PROCEDURES ARE REQUIRED.

SECTION 6 * REACTIVITY DATA

STABILITY: STABLE
CONDITIONS TO AVOID: OXYGEN, MOISTURE, POLYMERIZATION INITIATORS, COPPER, ALUMINUM
HAZARDOUS POLYMERIZATION: WILL NOT OCCUR (NORMAL COND)
CONDITIONS TO AVOID: NONE
INCOMPATIBILITY (MATERIALS TO AVOID):
OXYGEN, MOISTURE, POLYMERIZATION INITIATORS, COPPER, ALUMINUM
HAZARDOUS DECOMPOSITION PRODUCTS:
HYDROGEN CHLORIDE, CARBON MONOXIDE, AND POSSIBLE TRACES OF PHOSGENE

* * * 24-HOUR EMERGENCY ASSISTANCE: (304) 843-1300 * * *

FORM 6372 Rev. 3/88

Material Safety Data Sheet

PPG INDUSTRIES, INC.	ONE PPG PLACE	PITTSBURGH, PA 15272
* * VINYL CHLORIDE (VCM)		08/20/88 PAGE 4

SECTION 7 * SPILL OR LEAK PROCEDURES

STEPS TO BE TAKEN IF MATERIAL IS SPILLED OR RELEASED:
IMMEDIATELY EVACUATE AREA, REMOVE ALL SOURCES OF IGNITION AND PROVIDE MAXIMUM VENTILATION. ONLY PERSONNEL EQUIPPED WITH NIOSH/MSHA-APPROVED, SELF-CONTAINED BREATHING APPARATUS OR FULL FACE PIECE, AIRLINE RESPIRATORS WITH AUXILIARY SCBA'S OPERATED IN THE PRESSURE-DEMAND MODE AND SKIN/EYE PROTECTION SHOULD BE ALLOWED IN AREA. CLOSE OFF SOURCE OF LEAK AND ALLOW ALL SPILLED MATERIAL TO EVAPORATE. THEN WASH AREA OF SPILL WITH PLENTY OF SOAP AND WATER. CONTINUE TO THOROUGHLY VENTILATE AREA AND DO NOT ALLOW UNPROTECTED PERSONNEL TO ENTER AREA UNTIL MONOMER CONCENTRATION IS BELOW PERMISSIBLE EXPOSURE LIMIT.

WASTE DISPOSAL METHOD:
VINYL CHLORIDE EVAPORATES COMPLETELY, BUT ADEQUATE VENTILATION MUST BE PROVIDED TO PREVENT EXPLOSIVE MIXTURES. CARE MUST BE TAKEN WHEN USING OR DISPOSING OF CHEMICAL MATERIALS AND/OR THEIR CONTAINERS TO PREVENT ENVIRONMENTAL CONTAMINATION. IT IS YOUR DUTY TO DISPOSE OF THE CHEMICAL MATERIALS AND/OR THEIR CONTAINERS TO PREVENT ENVIRONMENTAL CONTAMINATION. IT IS YOUR DUTY TO DISPOSE OF THE CHEMICAL MATERIALS AND/OR THEIR CONTAINERS IN ACCORDANCE WITH THE CLEAN AIR ACT, THE CLEAN WATER ACT, THE RESOURCE CONSERVATION AND RECOVERY ACT, AS WELL AS ANY OTHER RELEVANT FEDERAL, STATE, OR LOCAL LAWS/REGULATIONS REGARDING DISPOSAL.

SECTION 8 * SPECIAL PROTECTION INFORMATION

RESPIRATORY PROTECTION:
NIOSH/MSHA-APPROVED, PRESSURE-DEMAND SUPPLIED AIR, RESPIRATORY PROTECTION WHEN AIR CONCENTRATIONS EXCEED THE PERMISSIBLE EXPOSURE LIMITS. SEE SECTION 7 FOR RESPIRATORY PROTECTION DURING SPILLS OR LEAKS. THE RESPIRATOR USE LIMITATIONS SPECIFIED BY NIOSH/MSHA OR THE MANUFACTURER MUST BE OBSERVED. RESPIRATORY PROTECTION PROGRAMS MUST MEET THE REQUIREMENTS OF 29 CFR 1910.134.

VENTILATION(TYPE):
USE GENERAL OR LOCAL EXHAUST VENTILATION SUFFICIENT TO LIMIT EMPLOYEE EXPOSURE BELOW PERMISSIBLE EXPOSURE LIMITS.

EYE PROTECTION:
CHEMICAL SAFETY GOGGLES

GLOVES:
NITRILE OR VITON(R) RUBBER GLOVES

OTHER PROTECTIVE EQUIPMENT:
BOOTS, APRONS, OR CHEMICAL SUITS SHOULD BE WORN WHEN NECESSARY TO PREVENT SKIN CONTACT. PERSONAL PROTECTIVE CLOTHING AND USE OF EQUIPMENT MUST BE IN ACCORDANCE WITH 29 CFR 1910.132 AND 29 CFR 1910.133.

SECTION 9 * SPECIAL PRECAUTIONS

PRECAUTIONS TO BE TAKEN DURING HANDLING AND STORING:
* THIS MATERIAL IS NORMALLY SHIPPED BULK IN TANK CARS AND BARGES. FOR DETAILS ON HANDLING PLEASE REFER TO THE PPG BROCHURE, VINYL CHLORIDE HANDLING AND PROPERTIES (A994-115).
* BEFORE UNLOADING, PURGE OXYGEN FROM UNLOADING SYSTEM.
* CONTAINER AND SYSTEM MUST BE ELECTRICALLY GROUNDED BEFORE UNLOADING.
* STORE AWAY FROM DIRECT SUNLIGHT AND OTHER SOURCES OF HEAT.
* AVOID CONTACT WITH FLAMES, HOT GLOWING SURFACES OR ELECTRIC ARCS.
* DO NOT USE IN POORLY VENTILATED OR CONFINED AREAS.
* STORE ONLY IN CLOSED, PROPERLY LABELED CONTAINERS.

OTHER PRECAUTIONS:
* KEEP OUT OF REACH OF CHILDREN.

---

* * * 24-HOUR EMERGENCY ASSISTANCE: (304) 843-1300 * * *

FORM 6372 Rev. 3/88

Material Safety Data Sheet

PPG INDUSTRIES, INC.	ONE PPG PLACE	PITTSBURGH, PA 15272

* * VINYL CHLORIDE (VCM)   08/20/88

* DO NOT BREATHE VAPORS. CANCER-CAUSING AGENT; VAPOR CAN ALSO CAUSE LUNG IRRITATION, DIZZINESS, ANESTHESIA, HELPLESSNESS OR UNCONSCIOUSNESS.
* AVOID CONTACT WITH EYES AND SKIN. CAN CAUSE FROSTBITE AND/OR TISSUE DAMAGE DUE TO RAPID EVAPORATION.
* DO NOT SWALLOW.
* DO NOT EAT, DRINK, OR SMOKE IN WORK AREAS.
* USE ONLY WITH ADEQUATE VENTILATION SUFFICIENT TO LIMIT EMPLOYEE EXPOSURE BELOW PERMISSIBLE EXPOSURE LIMIT.

COMMENTS:
SPECIAL FIREFIGHTING PROCEDURES (CON'T):

FOR SMALL FIRES, EXTINGUISH WITH CARBON DIOXIDE OR DRY CHEMICAL AGENTS AND WHILE CONTINUING TO COOL EQUIPMENT, ATTEMPT TO CLOSE OFF SOURCE OF LEAK.

THERE IS NO KNOWN EFFECTIVE EXTINGUISHING AGENT FOR LARGE FIRES. THEREFORE, IF SOURCE OF LEAK CANNOT BE CLOSED OFF, CONTINUE COOLING EQUIPMENT WITH STREAMS OF WATER TO AVOID TANK RUPTURE AND EXPLOSION AND ALLOW FIRE TO SELF-EXTINGUISH. NO ONE WITHOUT PROPER RESPIRATORY PROTECTION SHOULD ENTER AN AREA WHERE THERE HAS BEEN A FIRE UNTIL THE AREA HAS BEEN THOROUGHLY VENTILATED AND CHECKED TO BE SURE THE MONOMER CONCENTRATION IS BELOW THE PERMISSIBLE EXPOSURE LIMIT.

TSCA - VINYL CHLORIDE IS ON THE TSCA INVENTORY UNDER CAS NO. 75-01-4.

SARA TITLE III - A) LISTED IN 311/312 CATEGORIES AS FLAMMABLE, PRESSURE RELEASE, ACUTE AND CHRONIC, B) LISTED IN SECTION 313, C) NOT LISTED IN SECTION 302.

CERCLA - LISTED IN TABLE 302.4 AS A HAZARDOUS SUBSTANCE WITH A REPORTABLE QUANTITY OF 1 POUND. RELEASES TO AIR, LAND, OR WATER WHICH EXCEED THE RQ MUST BE REPORTED TO THE NATIONAL RESPONSE CENTER, 800-424-8802.

RCRA - WASTE VINYL CHLORIDE AND CONTAMINATED SOILS/MATERIALS FROM SPILL CLEANUP ARE U043 HAZARDOUS WASTE AS PER 40 CFR 261.33 AND MUST BE DISPOSED OF ACCORDINGLY UNDER RCRA.

CALIFORNIA PROP. 65 - THIS PRODUCT IS A CHEMICAL KNOWN TO THE STATE OF CALIFORNIA TO CAUSE CANCER.

R. KENNETH LEE
MANAGER, PRODUCT SAFETY

* * * 24-HOUR EMERGENCY ASSISTANCE: (304) 843-1300 * * *

FORM 6372 Rev. 3/88

20

Proper Use of the Laboratory in Assessing Occupational and Environmental Exposure to Toxic Chemicals

James C. Peterson, Ph.D.

Pacific Toxicology Laboratories, Los Angeles, CA

KEY TOPICS

    I. What test should be ordered?
   II. What type of specimen should be collected?
  III. When should the specimen be collected?
  IV. How often should specimens be collected?
   V. What are the background and toxic levels for the analyte?
  VI. Specimen archiving

Thanks to advances in analytical chemistry, trace amounts of various toxic chemicals can be analyzed accurately in environmental samples by commercial, governmental, and academic laboratories. A few labs even have the ability to measure these chemicals at low levels in human tissues. Having the ability to successfully perform such testing, however, does not necessarily mean that useful information will result. Analysis of human tissues requires an understanding of human biological processes not necessarily required for typical environmental analyses. Important considerations before requesting human toxicological testing are the following:

- What test should be ordered?
- What type of specimen should be collected?

- When should the specimen be collected?
- How often should specimens be collected?
- What are the background and toxic levels for the analyte?

While not all of these questions can be answered for each exposure situation; some attention must be given to each of them to assess any health risk.

I. WHAT TEST SHOULD BE ORDERED?

In any investigation of toxic chemical exposure, the chemical(s) suspected of causing the adverse health effect should be identified. The "shotgun" approach of testing for anything and everything is expensive and rarely results in useful data unless numerous negative results are of interest. The possible chemical(s) may be well-known to the individual through conscientious efforts to read container labels of materials used at work. If such information is not easily available, it may be obtained from Material Safety Data Sheets (MSDS) which the employer or manufacturer are required by law to supply to workers. If identification of possible chemical exposure is not this straightforward, analysis of environmental samples (wastes, air, water, soil, etc.) may be required. These tests can be prohibitively expensive for a private citizen.

In the workplace, both industrial hygiene air monitoring and wipe test results may be available to workers. If none of these sources of information is available, a realistic, logical evaluation of the individual's medical and job history, including any previous toxic exposures, (Chapters 5, 6, and 11) may suggest likely candidates for testing.

Any list of possible exposure chemicals must be narrowed further based on the availability of an appropriate test and the period of time which has passed since the presumed exposure. This will depend on the chemical and its metabolism in humans. In some cases it is not appropriate to measure the chemical itself (unchanged or as parent compound), since it may be extensively metabolized by the body immediately after absorption. Table 20–1 is a list of some commercially available tests for chemicals and their metabolites.

II. WHAT TYPE OF SPECIMEN SHOULD BE COLLECTED?

Only certain types of human specimens are appropriate for testing. Most commonly tested are urine, blood, and adipose tissue. Table 20–1 lists specimens appropriate to each chemical. Others, such as breast milk,

TABLE 20-1 Some Commercially Available Tests for Chemicals and Metabolites in Human Specimens

Chemical	General Population Range	Units	Toxic Level	Timing (half-life)	Specimen Requirements	
Solvents						
Benzene						
Benzene	<0.01	mg/l	N/A	H	7	ml LT
Phenol	<20	mg/g CR	50	H	25	ml U
Toluene						
Toluene	<0.01	mg/l	1	H	7	ml LT
Hippuric Acid	<1500	mg/g CR	2500	H	5	ml U
Cresol, o-	<0.1	mg/g CR	1	H	25	ml U
Xylene						
Xylenes	<0.01	mg/l	1.5	H	7	ml LT
Methylhippuric Acids	<50	mg/l	1500	H	5	ml U
Styrene						
Styrene	<0.01	mg/l	0.55	H	7	ml LT
Phenylglyoxylic Acid	<50	mg/l	250	H	5	ml U
Mandelic Acid	<50	mg/l	1000	H	5	ml U
Ethylbenzene						
Ethylbenzene	<0.01	mg/l	N/A	H	7	ml LT
Phenylglyoxylic Acid	<50	mg/l	N/A	H	5	ml U
Mandelic Acid	<50	mg/l	2000	H	5	ml U
Hexane						
Hexanedione,2,5-	<0.4	mg/l	5	H-D	15	ml U
Perchloroethylene						
Perchloroethylene	<0.003	mg/l	1	D	7	ml LT
Trichloracetic acid	<0.1	mg/l	N/A	D-W	5	ml U
Trichloroacetic Acid	<0.5	mg/l	7	D-W	7	ml LT
Trichloroethylene						
Trichloroethylene	<0.003	mg/l	N/A	H	7	ml LT
Trichloroacetic Acid	<0.1	mg/l	100	D-W	5	ml U

Substance					
Trichloroacetic Acid	<0.5	mg/l	N/A	D-W	7	ml LT
Trichloroethanol	<0.1	mg/l	N/A	H	5	ml U
Trichloroethanol	<0.1	mg/l	4	H	7	ml LT
Trichloroethane, 1,1,1-						
Trichloroethane, 1,1,1-	<0.1	mg/l	N/A	H-D	7	ml LT
Trichloroacetic Acid	<0.1	mg/l	10	D-W	5	ml U
Trichloroacetic Acid	<0.5	mg/l	N/A	D-W	7	ml LT
Trichloroethanol	<0.1	mg/l	30	H	5	ml U
Trichloroethanol	<0.1	mg/l	1	H	7	ml LT
Methylene Chloride						
Methylene Chloride	<0.003	mg/l	1	H	7	ml LT
Carboxyhemoglobin	<3%	%Hgb	5%	H	7	ml LT
Dimethylformamide						
N-Methylformamide	<5	mg/l	40	H	25	ml U
Petroleum Distallates						
Kerosene	<100	µg/l	N/A	H	7	ml LT
Mineral Spirits	<100	µg/l	N/A	H	7	ml LT
Ketones and Alcohols						
Methyl Ethyl Ketone	<0.01	mg/l	2	H	5	ml U
Methyl Ethyl Ketone	<0.01	mg/l	N/A	H	7	ml LT
Methyl Isobutyl Ketone	<0.01	mg/l	N/A	H	5	ml U
Methyl Isobutyl Ketone	<0.01	mg/l	N/A	H	7	ml LT
Methyl N-Butyl Ketone	<0.01	mg/l	N/A	H	5	ml U
Methyl N-Butyl Ketone	<0.01	mg/l	N/A	H	7	ml LT
Hexanedione, 2,5-	<0.4	mg/l	5	H-D	15	ml U
Acetone	<3	mg/l	20	H	15	ml U
Acetone	3–20	mg/l	N/A	H	7	ml GT
Isopropanol	<1	mg/l	N/A	H	4	ml
Methanol	<1	mg/l	15	H	7	ml GT

(Continued)

TABLE 20-1 (*Continued*)

Organochlorine Pesticides

DDT	<1.3	mg/kg	N/A	Y	0.5	gm AT
DDT	<20	µg/l	N/A	Y	5	ml S
DDE	<4.5	mg/kg	N/A	Y	0.5	gm AT
DDE	<50	µg/l	N/A	Y	5	ml S
DDD	<0.150	mg/kg	N/A	Y	0.5	gm AT
DDD	<20	µg/l	N/A	Y	5	ml S
Chlordane						
Chlordane, Alpha-	<0.01	mg/kg	N/A	W	0.5	gm AT
Chlordane, Alpha-	<0.5	µg/l	N/A	W	5	ml S
Chlordane, Gamma-	<0.01	mg/kg	N/A	W	0.5	gm AT
Chlordane, Gamma-	<0.5	µg/l	N/A	W	5	ml S
Oxychlordane	<0.150	mg/kg	N/A	Y	0.5	gm AT
Oxychlordane	<2	µg/l	N/A	Y	5	ml S
Trans-Nonachlor	<0.173	mg/kg	N/A	Y	0.5	gm AT
Trans-Nonachlor	<2	µg/l	N/A	Y	5	ml S
Heptachlor	<0.01	mg/kg	N/A	D-W	0.5	gm AT
Heptachlor	<0.5	µg/l	N/A	D-W	5	ml S
Heptachlor Epoxide	<0.130	mg/kg	N/A	Y	0.5	gm AT
Heptachlor Epoxide	<3	µg/l	N/A	Y	5	ml S
Miscellaneous OC Pesticides						
Lindane (gamma-BHC)	<0.030	mg/kg	N/A	D-W	0.5	gm AT
Lindane (gamma-BHC)	<13	µg/l	25	D-W	5	ml S
BHC, Beta-	<.470	mg/kg	N/A	Y	0.5	gm AT
BHC, Beta-	<10	µg/l	N/A	Y	5	ml S
Dieldrin	<0.180	mg/kg	N/A	Y	0.5	gm AT
Dieldrin	<4	µg/l	N/A	Y	5	ml S

Compound						
Endosulfan I and II	<0.5	µg/l	N/A	D-W	5	ml S
Endrin	<0.5	µg/l	N/A	D-W	5	ml S
Hexachlorobenzene	<0.360	mg/kg	N/A	Y	0.5	gm AT
Hexachlorobenzene	<5	µg/l	150	Y	5	ml S
Mirex	<0.043	mg/kg	N/A	Y	0.5	gm AT
Pentachlorophenol	<0.030	mg/l	2	W	15	ml U
Pentachlorophenol	<0.075	mg/l	5	W	5	ml PLSMA
Organophosphorus Pesticides						
Cholinesterase	Varies	IU	−30%	W-M	7	ml GN
Cholinesterase	Varies	IU	−30%	D-W	5	ml S
Alkylphosphate Metabolites	<0.03	mg/g CR	N/A	D	15	ml U
Malathion Metabolites	<10	µg/l	N/A	D	15	ml U
Parathion						
Para-Nitrophenol	<0.05	mg/g CR	2	D	15	ml U
Dursban						
Trichloropyridinol, 3,5,6	<0.01	mg/l	N/A	D	15	ml U
Herbicides						
2,4,5-T	<0.01	mg/l	N/A	D	15	ml U
2,4-D	<0.05	mg/l	N/A	D	15	ml U
Silvex	<0.01	mg/l	N/A	H-D	15	ml U
Polychlorinated Biphenyls						
PCBs Total	<20	µg/l	N/A	Y	4	ml S
PCBs Total	<50	mg/kg	N/A	Y	0.5	G AT
Polybrominated Biphenyls						
PBBs	<1.1	µg/l	N/A	Y	5	ml S
Anilines						
MBOCA	<10	µg/l	100	H-D	15	ml U
Methylenedianiline	<2.5	µg/g CR	N/A	H-D	15	ml U

(Continued)

TABLE 20-1 (Continued)

Heavy Metals						
Lead	<50	μg/CR	150	M-Y	15	ml U
Arsenic	<10	μg/g CR	50	H-D	15	ml U
Cadmium	<3	μg/l	10	Y	7	ml BT
Chromium	<1	μg/g CR	30	H-D	15	ml U
Nickel	<5	μg/g CR	35	D	15	ml U
Mercury	<20	μg/l	50	D-W	7	ml BT

μg = Micrograms
l = Liter
kg = Kilogram
g CR = Grams of Creatinine
DCL = Deciliter (100 ml)
H = Hours
D = Days
W = Weeks
M = Months
Y = Years
N/A = Not Available
AT = Adipose Tissue
U = Urine
S = Serum
PLSMA = Plasma
LT = Lavender Top Tube
GT = Green Top Tube
BT = Royal Blue Top Tube
mg = Milligrams
% Hgb = Percent Hemoglobin
g = Grams
IU = International Units

sebum (from sweat glands), hair and nails are analyzed less often, and therefore, data from such tests may be more difficult to interpret. While some chemicals can be measured in several different specimens, Table 20-1 lists the preferred specimens.

III. WHEN SHOULD THE SPECIMEN BE COLLECTED?

In most cases, the specimen should be collected as soon as possible after the exposure. Ideally, in workers, a baseline specimen, which can be analyzed or archived (frozen until needed), should be collected *before* the exposure. However, for chemicals which accumulate in the body, such as polychlorinated biphenyls (PCBs), timing of the analysis is not as critical as for solvents, such as perchloroethylene, which have short half-lives. Table 20-1 indicates the tests which should be done within hours (h), within days (d), within weeks (w), or within months to years (y) of exposure. While it may seem inappropriate to test for a chemical long after that chemical would be metabolized and excreted from the body, it may still be useful to test for its absence to convince a worried patient that the chemical has not been retained. A sensitive, negative test may be worthwhile as reassurance. This, of course, depends on the level of trust the patient has for accepted scientific and medical knowledge.

IV. HOW OFTEN SHOULD SPECIMENS BE COLLECTED?

This depends on the goals of the testing and the subsequent testing results. If the goal of such testing is to investigate an isolated exposure incident, a single sample may be sufficient. If chronic exposure is being monitored, then multiple samples would be necessary to establish the duration of exposure as well as the level of exposure. In general, shorter-lived chemicals (hours, days, weeks) should be tested more often than those which are retained for months or years.

V. WHAT ARE THE BACKGROUND AND TOXIC LEVELS FOR THE ANALYTE?

The background level or general population range refers to the level below which 95 percent of the values from non-occupationally-exposed individuals will fall. This is the chemical background. As more experience is gained with biological monitoring, background levels may change. Some chemicals, such as PCBs, are ubiquitous so that nearly all specimens will have measurable levels. Therefore, their mere presence in a specimen without a reference range provides the physician with very little new information, or worse, may lead to erroneous conclusions. The purpose of the general population range is to provide the doctor with a perspective from which to distinguish the typically exposed (background) from the atypically exposed. However, levels above or below this range do not imply toxicity or lack of toxicity. Table 20–1 lists general population ranges for commonly ordered tests. These ranges may vary according to the laboratory performing the test because they may be established from a variety of different databases. These general population ranges may be obtained from data published in the scientific literature or developed by the laboratory itself. Investigators involved in large studies often select an unexposed control group for comparison to the study population. This is particularly important when little data has been published concerning background levels of certain chemicals or when demographics may be an important factor.

The question of toxic levels is the most important and the most difficult question of all. The toxic level is the concentration of a chemical above which adverse health effects can occur. Since it is, of course, unethical to dose human beings with chemicals to determine the toxic level, most of our knowledge concerning a chemical's toxicity to humans is from retrospective epidemiological studies and accidental exposures. Some of these sources provide data with a high level of uncertainty, and actual toxic levels, especially from skin absorption, are often unavailable.

Currently, the most valuable source of guidelines for biological monitoring and limiting the levels of chemical exposure at work is the American Conference of Governmental Industrial Hygienists (ACGIH). This nongovernmental organization has developed guidelines which are called "Biological Exposure Indices" or BEIs. BEIs are levels of chemicals or their metabolites in various human specimens which should not be exceeded in order to avoid adverse health effects. These guidelines, with their recommended collection timing, are listed in Table 20–1. The ACGIH documentation should be consulted for proper application of the BEIs.

VI. SPECIMEN ARCHIVING

Specimen archival can be used for the following two purposes:

1. Specimens collected and frozen prior to employment or prior to a change in duties or working conditions can be used if needed at some time in the future to determine a baseline level for comparison to a post-exposure value. These specimens are especially valuable in cases

of accidental spills or the appearance of an adverse health effect potentially related to work.

2. Alternatively, specimens may be collected and frozen immediately after a chemical exposure incident and held until more information concerning testing is obtained.

There are several advantages to archiving samples:

- Analytical costs are reduced because only a portion of workers' specimens may need to be analyzed.
- When specimens are eventually analyzed, pre- and post-exposure samples are run together, eliminating day-to-day analytical variability and producing more precise and comparable data.
- Test methods, unavailable (even unforeseen) at the time of specimen collection, can be performed at a later date when they are developed.
- A specimen whose value depends on timely collection can be obtained immediately after an exposure. A decision as to which test should be performed (or whether a test should be performed at all) can be made later in a thoughtful, deliberate manner.

Archival of specimens (prior to potential exposure) is most effective for analytes which have the following characteristics:

- Long-term storage stability (months to years)
- Long biological half-life
- Blood or urine levels result specifically from chemical exposure and not from variable endogenous production or drugs
- Potentially present due to work environment and not to environmental exposure or domestic use unrelated to the job

Some analytes which meet these criteria are PCBs, PBBs, pentachlorophenol, organochlorine pesticides, and related compounds. Serum cholinesterase can be archived, but several baseline samples collected at different times should be stored. Urine trichloroacetic acid, phenol, hippuric acid, and other solvent metabolites can also be archived but should be analyzed within a shorter time frame (three to twelve months). Specimens can be analyzed for other analytes, but the successful use of archiving as a baseline measure depends on individual circumstances. The suitability of archiving specimens following an incident for future analysis depends on the stability of the analyte in the specimen. The holding time for analytes ranges from four days for methylene chloride in blood to five to ten years for PCBs.

Presently available tests, general population range (background), levels of concern, timing of analysis, and specimen requirements are shown in Table 20–1.

Part 5

Workers' Compensation Issues

21

Characteristics of 150 Work-Related Claims

Ilene R. Danse, M.D., and Linda G. Garb, M.D.

ENVIROMED Health Services, San Rafael, CA

KEY TOPICS

I. Catastrophes
    - Hazardous Chemical Waste
    - Heat Exposure
    - Poison Gas in a Hospital Toilet
    - Outcomes

II. Workplace Air
    - Exhaust, Smoke and Fumes
    - Gas Leaks or Contamination
    - Ammonia
    - Aldehydes
    - Acids/Corrosives/Reactive Chemicals
    - Solvents

III. Metals
    - Arsenic
    - Lead
    - Other Metals

IV. Plastics: Epoxies, Acrylics, Polyurethanes, Isocyanates, Hi-Tech
    - Multiple Symptoms
    - Asthma
    - Dermatitis
    - Catastrophes

V. Stress-Related Illnesses and Worry

VI.   Trauma/Orthopedic with Medical or Therapeutic Issues
VII.   Infections
VIII.   Cleaning Compounds: Disinfectants, Cleansers and Water
IX.   Dusts
X.   Pesticides
XI.   Office Environments
XII.   Drug Intoxication on the Job
XIII.   Summary of Findings in Work-Related Claims

These persons are part of a group of approximately 370 claimants examined in a clinical toxicology practice over several years. Claims considered not to be work-related are discussed in the next chapter. The cases are all real people; some details have been intentionally altered to ensure patient privacy. Generally, patients were referred because the diagnosis was uncertain or the degree of disability claimed was questioned. Some individuals were evaluated soon after a specific event. Others were evaluated years after leaving a workplace and, regardless of the initial incident, were frustrated and angry at being embroiled in so lengthy a workers' compensation evaluation process. Some persons lost their homes, cars and trucks, or died of unrelated causes before their claims were resolved.

Our findings in work-related claims are as follows. One-hundred-fifty individuals were found to have work-related aspects to their claims as listed in Table 21 – 1. As few people used only one substance, they were catego-

TABLE 21-1   Findings in 150 Work-Related Claims

Circumstances	No.	Percent
Serious/catastrophic	6	4.0
Air	38	25.3
Metals	30	20.0
Epoxies/plastics/high-tech	21	14.0
Stress/worry	17	11.3
Trauma with medical or therapeutic issues	15	10.0
Infections	6	4.0
Disinfectants, cleansers, water	4	2.7
Dusts	4	2.7
Pesticides	3	2.0
Office	2	1.3
Drugs	2	1.3
Miscellaneous	2	1.3
TOTAL	150	100

rized based upon the suspected culprit. In this chapter, "permanent disabilities" are defined as an injury that would be compensable and rated as such.

I. CATASTROPHES

Eight persons involved with serious or catastrophic incidents were in this group; however, two of them will be discussed in the section covering plastics. Catastrophes involved truly hazardous wastes, heat stress and chemical warfare gas generated in a hospital toilet from chlorine bleach and ammonia.

Hazardous Chemical Waste

Three cases were referred to us by different sources, but coincidentally, all involved the same exposure.

> Two hazardous waste haulers took appropriate protective measures for loading a liquid waste, but a different substance was actually loaded. They became ill immediately; symptoms were protracted.

> By coincidence, a reactor operator from the plant that generated the waste was evaluated because of symptoms persisting one year following a reactor upset during which he wet his skin and inhaled fumes. He did not appear to know of the other two men. The waste from the reactor upset may have been loaded by the two haulers.

> No factors for secondary gain could be identified. The three men, all of whom were "macho," were embarrassed by their remarkably similar symptoms, consisting of random fainting attacks, loss of potency, unpredictable chest pains simulating myocardial infarctions, and "pins and needles" sensations (dysesthesias). All had evidence of autonomic nervous system instability (including fainting, irritability and depression) more than one year after the exposure.

As it turned out, the waste was from a product manufactured for export. Reports of spills and misadventures from the country of import revealed other cases of long-lasting autonomic nervous system dysfunction.

Heat Exposure

There were two cases of excessive heat exposure with loss of consciousness in both cases. In one case, heat exposure plus lack of oxygen resulted in an anoxic injury and a close brush with death.

A construction worker passed out in a heated, airless space enclosed in plastic sheets. At first he had a seizure disorder, then he developed addiction to headache medication.

A case of heat stroke was caused by *inappropriate use of protective clothing.*

Reportedly because of a work rule, a vegetable picker wore a rubber suit in weather over 100 degrees. The worker complained to the field supervisor, who would not relax the rule *nor* allow the worker to leave the field until after she became quite ill. When hospitalized, her body temperature was 108 degrees. Initially she was delirious, but she recovered fully *without* permanent impairment.

Poison Gas in a Hospital Toilet

One of the saddest cases in the series involved a hospital toilet. It seems that in hospitals and health care institutions, settings in which the risk of physical harm is great, mean-spirited politics is prevalent. An extreme case is described below, but the mentality was widespread.

A man who worked as a dispensary clerk had two other jobs and also went to school. He had a pre-existing history of asthma but had sustained no disability or lost time from work because of it. When he became an employee in the dispensary, he was a union steward at a time when other dispensary employees were non-union. The dispensary was a small, enclosed room with a glass window through which the clerk processed drug orders; inside, behind him, was a small bathroom for employees.

Daily, during this employee's shift, a housekeeper poured chlorine bleach, then ammonia into the toilet bowl, generating annoying fumes into an enclosed space. Repeated requests to housekeepers to cease and desist were ignored. Labels were also ignored—every bottle of chlorine bleach is labeled, "do not mix with ammonia," and every bottle of ammonia bears "do not mix with chlorine bleach." On more than one occasion, several pharmacy employees, as well as those located one floor above, went to the emergency room because of symptoms from fumes. Emergency room physicians from this very hospital, and employees' personal doctors, asked that ammonia and chlorine not be mixed. As several requests from both physicians and employees, including some in writing, were ignored, eventually (and reluctantly) OSHA was called. The OSHA inspector removed the bottles of bleach and ammonia from the cart. A citation was given, along with a warning *never* to use these materials in combination again.

After two or three abstemious days, the noxious mixture again appeared in the toilet, which, by the way, was used only by two or three dispensary employees who were long past wanting to nail it shut. The worker, who evidently needed or wanted this job very badly, was hospitalized after a few more weeks in "status asthmaticus," and nearly died. After a stormy hospital course, he was eventually discharged, but he sustained extremely severe permanent lung injury; various parameters of lung function were between 3 percent and 10 percent of predicted values. This worker, who had been a vigorous and dynamic individual, is now a pulmonary cripple.

It is well-known that chlorine bleach and ammonia, when mixed together, generate chloramine, a poison "chemical-warfare" gas that permanently destroys lung tissue. Literature reported lung damage in housewives who had mixed chlorine bleach and ammonia together during spring cleaning. With brief and less intense exposures, all cases sustained permanent lung damage.

Outcomes

Of the six persons discussed in this category, four were medically unable to return to work. Two were working but at different occupations; it was anticipated that a third would return to the work force. At least two cases were preventable. Both were caused by inappropriate use of safe products —the protective suit and the cleaning compounds.

II. WORKPLACE AIR

Thirty-eight claims (25.3 percent) were submitted because of problems in workplace air. The substances or issues provoking these complaints were the following:

- motor or engine exhaust, smoke or fumes entering indoor environments
- gas leaks or contamination
- ammonia leaks
- aldehydes
- acids and corrosive or reactive chemicals
- solvents

Exhausts, Smoke, and Fumes

Six cases were examined because fumes from motor or engine exhausts were generated indoors, either routinely or because of faulty motors in indoor equipment. In other cases, truck engine exhausts or incinerator smoke backed up into people's indoor work areas. These persons wanted to be examined by a doctor or have the situations corrected; there were no injuries in this group.

Gas Leaks or Contamination

The substances of concern were freon-contaminated propane fuel and mercaptans. A group of restaurant workers were inadvertently exposed to freon-contaminated propane fuel believed to generate phosgene gas when burned during cooking. Workers began to wonder whether their own lungs were damaged when they noticed that shiny metal surfaces in the restaurant kitchen had suddenly become corroded. None of them lost time from work until the restaurant was ordered to close. Two workers who also smoked showed decreased diffusion capacity of the lung that seemed out of proportion to smoking history.

Mercaptans, naturally present in human gas, are added to odorless propane or natural gas, for two reasons. They have a low toxicity but strong sulfurous odors which provide early warning and facilitate detection of gas leaks. Several cases involved complaints of security or law-enforcement persons investigating gas leaks.

Ammonia

Ammonia or ammonium salts accounted for complaints in six individuals unrelated to each other. The case of lung damage from chloramine gas, described in the section on catastrophic illness (see page 178), has been considered separately. Persons who encountered ammonia compounds were janitorial workers, lab technicians, beauticians, refrigeration workers, and firefighters.

In addition to symptoms of mucous membrane irritation, which would be expected, two persons developed symptoms that were not recognized as being due to ammonium salts. In two unrelated incidents, "non-toxic" fire extinguishers were accidentally released in enclosed spaces. Both persons developed a delayed reaction consisting of chest pains so severe that each was hospitalized as a possible heart attack victim. Once the chest pains subsided, there was no residual impairment.

Aldehydes

Aldehydes prompted complaints by hospital technicians forced to work with formaldehyde or glutaraldehyde in small, enclosed areas without appropriate ventilation. These employees insisted that their complaints were repeatedly ignored by hospital superiors, compelling them to submit workers' compensation claims so that the employer would take notice and implement appropriate safety procedures.

Acids/Corrosives/Reactive Chemicals

Injuries ranged from mucous membrane irritation to lung damage as a result of exposures to acid mists, fluoride salts, chlorine, concentrated sufur dioxide, and hydrogen sulfide. In most instances, there was only transient mucous membrane irritation that disappeared once work practices were improved or the person was removed from the workplace. However, a food processing worker who had breathed in sulfuric acid mists for years without respiratory protection had significant loss of lung tissue but was still working at that job.

Solvents

Six complaints were primarily due to solvents in the air. Many other workers mentioned solvents or were concerned about using them, but they are categorized elsewhere according to their main exposure — for example, epoxy resins, paints, or wood dusts. In this group — consisting of six individuals working in printing, electronics and fabricating machinery — the solvents methanol, methyl ethyl ketone, toluene, butyl acetate, and acetone were believed to be related to the symptoms, mainly mucous membrane irritation, as cause-and-effect. Three of the six individuals used large quantities of solvent in small, enclosed, poorly-ventilated work areas and became temporarily dizzy. None were felt to have any organic brain damage as a result of some solvent use at work.

In five out of six complaints, symptoms were transient, temporary and corrected by good work practices. One printer appeared to have a relatively high exposure to solvents in the workplace as well as an alcohol abuse problem and also smoked heavily. In addition to mucous membrane irritation, he appeared to have memory loss, headaches, and cardiac arrhythmias.

Thus, out of 38 complaints generated by legitimate questions about the quality of workplace air, only *four* persons had evidence of permanent lung disablement at the time of the evaluation. In all four, the diffusing capacity of the lung was altered, and the exposures were to noxious materials such as phosgene, sulfuric acid mists, or sulfur dioxide. Only *one* person out of this group, a trucker who had a flu-like illness from sulfur dioxide after (dip) "sticking a tank," was ill enough to lose time from work. Three other persons (two with ammonium sulfamate exposures from fire extinguishers and the printer with arrhythmias) were hospitalized but returned to work without evidence of permanent disablement.

III. METALS

Forty-one claims were initiated, all or in part, because of possible exposures to metals present in the workplace. In approximately 30 cases, or 20 percent, metal exposure was the main reason for the evaluation. Of the 30 cases, three had permanent disabilities—sinusitis, dermatitis, and skin cancer. None of these precluded a return to work at the same job, but work practice improvements were recommended.

Arsenic

Arsenic was the metal of concern in 15 cases. Seven people were part of one incident and were referred both for a comprehensive evaluation and to "put it on the record." Another individual who worked with arsenic had hyperkeratosis. One person presented with multiple skin cancers that were felt to be work-related, with arsenic being one of several substances associated with skin cancers. The remainder worked with or near arsenic or arsine gas in the semiconductor industry. The only patient with skin cancers had them removed and was working for the same employer at the time of the evaluation.

Lead

Ten referrals involved possible lead exposures. Five persons were referred because lead levels were elevated and three because of symptoms. They needed advice on treatment, disability, and medical clearance for return to work. OSHA regulations preclude returning a worker to a workplace until two consecutive blood tests indicate the blood lead level is at or below 40 μg/dL. Lead-exposed workers were painters engaged in sanding or burning lead paint, radiator repairman who used lead solder, and lead fishing sinker manufacturers. All smoked and ate while working, and none were aware of any laws, guidelines, or protective practices that prevented lead exposure.

Only one asymptomatic person with an elevated blood lead level intended to return to his original profession; the others were embroiled in litigation and furious at having been "poisoned." While the claims in this group were legitimately job-related, and some persons were temporarily disabled, from a medical and toxicological standpoint, all could have returned to their jobs with straightforward implementation of hygienic work practices and minor modifications.

Our experience with these angry people underscored a clear need for prompt evaluation, education, and reassurance!

Other Metals

Other metals included chromates and metal plating operations with presentations of sinusitis, dermatitis, or metal fume fever.

IV. PLASTICS: EPOXIES, ACRYLICS, POLYURETHANES, ISOCYANATES, HI-TECH

Multiple Symptoms

At least 21 persons were referred because of potentially serious reactions as a result of using uncured plastics. (There were a number of secondary complaints from these same materials, but the workers claims are listed in other categories such as "solvents," which were felt to be the primary cause of injury.) Workers were spray or industrial painters using epoxy paints or polyurethane coatings. They worked in the high-tech, electronics, or defense industries, auto painting, or in the manufacture of synthetic rubber or plastics. Their jobs were in the areas of electronics, coatings, plastics manufacture, rubber production, shipyards, and in arts and crafts, assembly and extrusion. Offending substances consisted of epoxies, acrylonitrile, polyurethanes; isocyanates including toluene diisocyanate, hexamethylene diisocyanate and polyisocyanates; acrylamide resins, methacrylates, phenol formaldehyde resins, and diphenylamines. In several instances, the content of the compounds used could not be obtained from the workplace.

In general, this group was otherwise healthy, had a low incidence of alcohol and cigarette use, and, once the substances were known, diagnoses were straightforward. Complaints involving several organ systems were frequently observed, including headaches, wheezing, skin reactions, and flu-like symptoms, as well as dermatitis. Some case examples follow:

A 39-year-old foreman reconditioned a 30-foot-long tank. By the second work day, he had myalgias (muscle aches), vomiting, sleeplessness, chills, diarrhea, a feeling of impending doom and blurred vision. He had developed a makeshift oven to "bake out" a tank that was too long to fit in standard ovens. Phenol formaldehyde resin, used without proper exhaust ventilation, was the likely offending agent. The patient recalled that once before when he had used the same material, also under makeshift conditions, he had developed a similar flu-like reaction.

Another man, a 32-year-old boat repairman and spray painter, worked at the same job for nine years. However, he developed central nervous system, respiratory, and skin complaints following the introduction of a new line of epoxy paints, diphenylamine, polyacrylamide resins, and newer solvents.

A 52-year-old dental technician repaired and fabricated dental prostheses using acrylic materials for 16 years. He repeatedly asked his employer, a dentist, if he could have a fume hood which cost $100, but the employer refused (!). After eight years, this man developed progressively more severe headaches, unrelieved by six to eight potent (and habit-forming) codeine-containing analgesics a day, and nasal sprays. After 14 years at the job, he developed wheezing which was relieved somewhat by steroids. After 16 years, he was at his wit's end. He would leave work with a combination of incapacitating headaches he thought were from the acrylics at work and from the medication he was taking for them. A neurologist temporarily removed him from work to see how he would fare. After a week at home, the headaches completely disappeared. By the second week, he required no medication of any kind, and felt well. On his own, he decided to re-challenge himself. He had a little room where he liked to sit in a comfortable chair and read a newspaper. He closed the windows, shut the door, and from acrylic powder that he had brought home from work, mixed up a batch of acrylic resin and put it on a little side table next to him. Within 30 minutes, he again had an incapacitating headache.

Several persons experienced flu-like reactions to paints and coatings, all on more than one occasion. The workers themselves realized that every time they used a particular mixture, they would develop a reaction.

Asthma

Isocyanates, polyurethanes, and diamines appear to be the putative agents for most of our cases of industrial asthma. Seven persons with asthma due to isocyanates could not return to jobs requiring use of the offending material, but there was no disablement other than a work preclusion against using isocyanates. (A work preclusion is not a compensable injury.) In other cases with asthma, the employer was cooperative, and the use of epoxies or offending materials was not essential, so that, for instance, wire replaced the use of epoxy or an ink that contained ethylene diamine was replaced with another ink.

Dermatitis

The incidence of dermatitis in workers using uncured plastics was high. Skin reactions included welts and redness, roughening and inflammation (dermatitis). In some cases dermatitis occurred because the best, busiest, and fastest line workers apparently were getting epoxies or other substances on their skin and not taking time to remove it.

Workers with dermatitis and their employers were quite cooperative.

Workers were more than willing to change to more hygienic work practices in order to stay on the job. Likewise, employers offered to substitute other materials, provide coveralls and other skin protectors, or provide tools that avoided skin contact, etc. They were even willing to make process changes in order to keep these workers. As a result, workers who improved their personal hygiene, used protective equipment, or modified their work practices, were able to stay on the job.

Three persons with severe skin reactions were unable to return to a job requiring the use of epoxy. One of them was unable to work at all because of poor skin condition.

Catastrophes

Two people were injured in catastrophes. In one, an explosion was followed by fire. The worker sustained burns and smoke inhalation, fortunately there was no permanent disablement, as both injuries healed.

In another incident, a reactor operator manufacturing acrylamide gel was drenched by acrylamide in chloroform when a valve ruptured. At the time, he wore a respirator, promptly showered and changed his clothes. However, he developed severe depression, muscle weakness, and muscle wasting of the lower extremities similar to that observed in animal experiments and was still incapacitated two years afterward.

As a result of these cases, we would certainly recommend that these materials be used with safeguards in enclosed systems.

V. STRESS-RELATED ILLNESSES AND WORRY

Persons with psychiatric claims were referred to the appropriate specialist. The patients described here were referred for an internal medicine/toxicology evaluation because of a component of stress appropriate to these specialties. Seventeen persons (11.3 percent) claimed stress at work aggravated medical conditions or feared poisoning from a toxic exposure, such as fear of hazardous wastes.

> A hospital kitchen worker was distraught because incinerator fumes from contaminated plastic hospital wastes continually backed up into his work area whenever the incinerator was used.

Some claims were from firefighters who unwittingly had put out fires in areas later found to contain substances with bad reputations or listed as hazardous.

In one incident, firefighters returned from several days off duty to find that their equipment had been decontaminated but that they had not been notified of a possible exposure to polychlorinated biphenyls (PCBs), nor had provisions for medical checkups been arranged. At least one disability retirement resulted from this incident.

The main issue in all these cases appeared to be that their superiors did not care for them enough as human beings.

Apart from hazardous wastes, other persons claimed that stress-related diseases such as hypertension, ulcers, and colitis were aggravated by their jobs. Several were white-collar workers, but the majority of workers in this group were in public service, such as highway patrol or police work, prison work, and firefighting. For some public servants, regulations presumed that certain disorders, such as heart conditions, were job-related unless rebutted by evidence.

A source of stress in law enforcement and prison workers was the possibility of being bitten or splashed with excrement by HIV-positive persons, or persons with AIDS.

Workplace politics, psychological harassment, lack of human kindness, and mean-spiritedness were issues that were frequently raised in the histories obtained from these patients. Color, sex, and sexual orientation were contentious issues between coworkers, and between supervisors and employees. Generally, speaking, they were medically capable of doing their work and were undergoing psychiatric evaluations.

VI. TRAUMA/ORTHOPEDIC WITH MEDICAL OR THERAPEUTIC ISSUES

In 15 cases (10 percent), an initial claim of trauma, repetitive strain, or lifting injury was made—including one from lifting a cow. ("Why did you lift a cow?" "He fell over.") These patients had a medical complication or a complication of treatment for the original injury. Several developed thrombophlebitis or cellulitis. Several had peptic ulcers or other drug-induced illnesses, such as drug-induced lupus which resulted from nonsteroidal anti-inflammatory drug treatment for an industrial injury. Complications of treatment or medication side-effects frequently were not recognized by treating or evaluating physicians.

In our opinion, several persons had received inappropriate treatment.

An individual with a major depression stayed home and cried for five years. Treatment for depression was a lawyer's referral to a psychologist. No other treatment or consultation was suggested.

A woman had a surgical procedure for an industrial injury, then was completely disabled by tardive dyskinesia that had gone unrecognized for several years. No one considered the relation between her disabling neurologic symptoms and the phenothiazine medication that she was taking.

In this group as a whole, many persons smoked and used alcohol regularly, and generally, following an injury, the consumption increased. However, from a strictly medical standpoint, only three of these individuals could not resume their usual and customary occupations; of these, one was working at another job.

VII. INFECTIONS

Six persons were considered to have work-related infections. Of two cases of brucellosis with permanent impairment, one man became a pig farmer.

Two butchers (in different counties) developed brucellosis. Neither case was promptly diagnosed or treated, and both went on to develop chronic syndromes of brucellosis asthenia (extreme fatigue).

A heavy equipment operator worked on a construction job in an area where Valley Fever (coccidioidomycosis) was endemic. Since he lived elsewhere, his case of coccidioidomycosis was considered to be industrial.

Chorioretinitis developed in a utility worker when he got untreated sewage sludge in his eye.

Two cases of hepatitis were observed in firefighters rendering medical assistance and cardiopulmonary resuscitation (CPR). One of them developed acute fulminant hepatitis, accounting for the only death in this series. This officer had performed mouth-to-mouth resuscitation on an infant thrown from a car in a vehicular accident; the child's mother was a drug-abuser.

VIII. CLEANING COMPOUNDS: DISINFECTANTS, CLEANSERS, AND WATER

Three hospital and convalescent home housekeepers complained of symptoms caused by disinfectants. Two of them usually had wet hands, but they did not wear gloves; the employers neither insisted upon them nor provided them. Consequently, they became permanently disabled with chronic dermatitis. One person who wore gloves complained of conjunctivitis.

One case was due to water alone.

A 30-year-old man was team leader for a centralized kitchen where dough was washed from large baking trays. A floor drain did not work well, and he usually found himself sloshing in puddles of water. He asked the employer for rubber boots, but, according to the worker, the employer refused. His shoes and feet were constantly wet. Painful ulcers and itching developed on his feet evolving into a massive fungal infection, and the skin sloughed. A dermatologist removed him from work; he was in a wheelchair for six months. Finally, the deep ulcers on his feet healed and were filled in by fresh, new pink skin. He was returned to work with a preclusion to "stay off his feet," so he was put on the assembly line, where he could sit. But, he was very short; in order to handle the dough on the line, he needed a higher stool. The employer refused.

It is likely that in a case of this type, the worker will undergo vocational retraining. He has generated more than six months of doctor's bills for treatment, bills for various evaluations, legal fees, and wages for lost work time. This case could cost $70,000 or more. To think that this man's suffering and indignity, to say nothing of the expense, could have been avoided if the employer had provided a pair of rubber boots!

IX. DUSTS

Four claims were submitted because of dusts. Two cabinet makers developed asthma or hyperreactive airways, in response to wood dust.

A case of intractable asthma developed in an agricultural worker who processed and packaged beans. Bean dust was felt to be the allergen.

An intractable dermatitis was aggravated by uncured cement dust.

X. PESTICIDES

Three persons had work-related pesticide exposures. In none of them was there permanent injury or inability to return to their usual and customary profession.

One woman had asthma and spasm of her throat after pyrethrins were applied to her work area in a food processing plant.

One pesticide applicator drenched himself with an organophosphate cholinesterase-inhibiting pesticide which was absorbed through the skin. He was hospitalized with organophosphate poisoning. When he returned to work within a week, he found that he had been replaced.

One nurseryman who applied multiple fertilizers and pesticides developed conjunctivitis that probably was work-related.

XI. OFFICE ENVIRONMENTS

Office environments are not immune. Dermatitis of the hands may develop due to carbonless copy paper and photocopy or FAX paper. At least one of these cases has required change of jobs. One person has a work preclusion to avoid carbonless copy paper.

XII. DRUG INTOXICATION ON THE JOB

Law enforcement officers were accidentally intoxicated with illicit drugs in the course of duty.

A highway patrolman became permanently disabled in a freak accident involving LSD, while learning to identify confiscated drugs "in the field."

A narcotics officer was intentionally splashed with phencyclidine (PCP), ether and an unknown liquid during a "drug bust." Some PCP and ether were absorbed through the skin.

XIII. SUMMARY OF FINDINGS IN WORK-RELATED CLAIMS

In 150 persons who were evaluated because of complaints felt to be arising from the workplace, there were some serious incidents. There was one death of a firefighter resulting from fulminant hepatitis acquired in the course of CPR.

Overall, this group was characterized by credible symptoms from real exposures; there were no discoveries of a brand new "amazing toxic reaction."

The largest single group of compounds causing problems was in the category of uncured plastic resins and coatings. In some cases, workers diagnosed their own problems in relation to the offending agent, but their employers did not believe them. In other cases, employers were more than willing to alter work practices so the employee could return. Persons with asthma due to isocyanates could not return to work areas containing isocyanates. All but one of them were working or undergoing vocational retraining at the time of the evaluation and were asymptomatic away from isocyanates.

Few people were so disabled that they could not work again, although long-term follow-ups were not available. Out of 370 claims, fewer than two

TABLE 21-2 Permanent Disabilities

Category	Substance	No.	Comment	T.D.[a]	Able to Work? Yes/No	Able to Work? Same Job?
Catastrophe	Truly hazardous waste	3	3 Autonomic nervous system dysfunction	2	1 Yes	3 No
Air	Chlorine bleach and ammonia	1	1 Prolonged lung damage	1	No	No
	Phosgene	2	2 Decreased diffusing capacity		2 Yes	Yes
Metals	Sulfuric acid fumes	1	1 Lung damage		Yes	Yes
	Arsenic	1	1 Skin cancer		Yes	Yes
Plastics	Isocyanates, polyurethanes, diamines	7	7 Asthma		7 Yes	No
	Epoxies	3	3 Dermatitis	1	2 Yes	3 No
	Acrylamide	1	1 Peripheral neuropathy	1	No	No
Toxic/Stress	"Hazardous waste"	1	1 Stress/disability retirement[b]		Na[c]	No
Therapeutic	Disabling side-effect or inappropriate treatment	3	1 Drug-induced lupus	1	NA	No
			1 Untreated depression		NA	No
			1 Tardive dyskinesia		No	No
Infections	Butchers	3	2 Brucellosis	3	1 Yes	2 No
	Firefighter (EMT)		1 Hepatitis		No (died)	
Cleaning	Water	1	1 Dermatitis		NA	No
	Disinfectant	1	1 Dermatitis		Yes	Yes
Drug intoxication	Law enforcement	2	1 LSD	1	2 Yes	1 No
			1 PCP			

a T.D. = Totally disabled from usual and customary occupation.
b. Psychiatric.
c. NA = not available.

190

dozen persons had a permanent disability that was ratable as defined by workers' compensation. At the time these persons were evaluated, about 11 people (including the death and a psychiatric claim) were permanently disabled from working at their usual and customary occupation. Eight of them probably could not return to work at any job. The status of those workers is listed in Table 21–2.

In at least 95 percent of the work-related claims, the claimants were either working or involved in a retraining process.

22

Findings in 220 Non-Work-Related Cases

Ilene R. Danse, M.D. and Linda G. Garb, M.D.
ENVIROMED Health Services, San Rafael, CA

KEY TOPICS

One hundred fifty persons believed to have work-related complaints were described in the preceding chapter. In this chapter, 220 persons with complaints that were unrelated to work are discussed. These cases are tabulated in Table 22-1.

I. DRUG EFFECTS

Drug symptoms comprised the largest group of claims. Forty-six persons, approximately 21 percent of this group, had symptoms related to substance abuse or prescription drugs. A detailed breakdown of this group is presented in Table 22-2, which includes two cases of lung cancer in smokers.

Substance Abuse

Most claimants were substance abusers who presented because of symptoms related to tobacco, alcohol, or illegal drugs; thirty-one people were in this group. Substance abusers frequently feigned industrial injuries in order to obtain drugs legally.

> A health professional "injured" his back when turning a patient. Frequently, after lifting or moving a patient, he visited his doctor, who prescribed habit-forming medication. He abused both legal and illicit drugs in increasing quantities. As drug-seeking behavior escalated, he became totally unable to work.

Prescription Medication Effects

Fifteen persons presented with side-effects of prescription drugs or addiction to prescription medication as the major diagnosis.

Prescription drug side-effects. Workers in this group were taking medications with side-effects compatible with their symptoms. However, this obvious source of overexposure to "toxins" had not been recognized by treating physicians or patients. Some persons, innocently unaware that their symptoms were drug side effects, were often treated by their doctors with more medication, not removal.

Prescription drug addiction. Addiction to prescription drugs—often narcotics, sedatives, pain medications, and "anti-anxiety" drugs—was a frequent problem.

> Carried to the extreme, a patient who visited a doctor for a possible work-related complaint was given prescriptions which, over a two-year period,

TABLE 22-1   Findings in 220 Cases Found Not to be Work-Related

No.	Percent[a]	Findings	Comments
46	20.9	Drug effects	See Table 22-2
41	18.6	Normal	No illness
24	10.9	Allergic rhinitis (hay fever) or mucus	No disability
16	7.2	Viral (other than respiratory)	Shingles, post-polio syndrome, labyrinthitis, chalazion, Guillain-Barre Syndrome, viral hepatitis, pancytopenia following influenza B, Acquired Immune Defficiency Syndrome (AIDS)
14	6.3	Cardiovascular	Arteriosclerotic coronary artery disease, congestive heart failure, mitral valve prolapse, hypertension and hyperlipidemia
12	5.5	Respiratory	Bronchitis, sinusitis, coughs, colds, sore throat and flu
11	5.0	Sexual	Pregnancy, ectopic pregnancy, pubic lice, chlamydia, miscarriages, sex-change operation side-effects, uterine fibroids, breast lump
10	4.5	Nervous system	Migraine headache, CVA, Bell's Palsy, Parkinson's Disease, petit mal epilepsy
9	4.1	Asthma	Neither work-caused nor aggravated.
7	3.2	Dermal	Irritation from dentures, vitiligo, contact dermatitis, self-inflicted burns, boot-blisters, tinea pedis (Athlete's foot), acne rosacea
6	2.7	Connective tissue disorders	Degenerataive arthritis, rheumatoid arthritis, sarcoid, lupus, monoclonal gammopathy with peripheral neuropathy, dermatomyositis
5	2.3	Metabolic/endocrine	Diabetes, porphyria, thyroid nodule, Graves' disease, sleep apnea
4	1.8	Other infections	Epididymitis, tinea pedis, coccidioidomycosis, tuberculosis
4	1.8	Renal	End-stage kidney failure, microscopic hematuria
11	5.0	Other	Congenital clubbing, Von-Recklinghausen's Disease, food allergies, gastric ulcer, iron deficiency anemia, psychosis, spurious, Baron Munchausen's Syndrome

a.   Percentages have been rounded off and, therefore, do not add up to 100 percent.

TABLE 22-2   Non-Work-Related Drug Effects in 46 Persons

No.	Effect and (Subtotal)
31	Substance Abuse
	(11) COPD[a] in smokers
	(10) Complications, due to alcohol abuse
	(8) Complications, due to illicit drugs
	(2) Lung cancers in smokers
15	Prescription Medication: side-effects or addiction

a. COPD—Chronic obstructive pulmonary disease

escalated to about 1,000 pills a month. The person claimed to be unable to work—with that number of pills to take, when would there be time?

Persons prescribed (and taking) hundreds of pills a month were not unusual.

While this group claimed injuries, objective findings apart from symptoms of drug overdosage were unusual. "Drug-seeking" behavior was the rule, rather than the exception. Some persons (and some doctors) feigned work-related injuries, such as back or neck sprains, to justify drug prescriptions.

An hourly-rate temporary employee worked on the line boxing Christmas gift items for approximately seven weeks, when the job ended because the work was done. Some time thereafter, the person developed "elbow pain." Medical records indicated that this individual was placed on narcotics for the "elbow pain" and for five years had received escalating amounts of codeine-type narcotics. When examined, this individual was so spaced-out, she was barely able to walk.

An individual had disc surgery, but it was questionable that the surgery was indicated. In any event, $32,000 of narcotics was prescribed for subsequent back pain. The patient was suspected of selling some of his medicine "on the street."

Of persons examined, abuse of "legal" prescription drugs was more frequent than illicit drug abuse (Table 22–2). Invariably, drug bills were paid for by insurance.

Substance Abuse Fatalities

Table 22–2 summarizs people who were examined. Substance abuse fatalities or totally disabling accidents are not tabulated because patients were not examined, rather, charts and autopsy records were reviewed. However, these cases are worth mentioning because fatal or catastrophic accidents occurred during alleged or actual work-related activities. In these cases of death or quadriplegia, in*toxic*ating quantities of alcohol and "recreational" drugs were found.

II. NORMAL PEOPLE

This group was normal by all parameters — physical examination, laboratory data, and evaluation of medical files. Persons who were normal or had drug reactions frequently stated that "my doctor doesn't know what is wrong with me; therefore, it must be due to the job."

This group displayed a need for information or an exam and were concerned about statements on Material Safety Data Sheets, such as "carcinogens" and "toxic" substances. Some felt that submission of a workers' compensation claim would document work with "carcinogens" in case they later developed cancer. People lacked clear information from their physician, place of work, or their union. Those who had received information did not understand it.

III. ALLERGIC RHINITIS/MUCUS

This group had dramatic overreactions to trivial complaints, such as noticing a speck of blood when they blew their noses or having a post-nasal drip. While not disabling, annoying hay fever-type symptoms consisting of some or all of the following were widespread: watery eyes, runny nose, throbbing swollen sinuses, tightness or heaviness in the throat or chest. Twenty-four persons, or about 11 percent, had allergic rhinitis or some complaint related to hay fever, "allergies," or mucus.

Some "mucus complainers" attributed their woes to poor-quality air in the building or blamed chemicals in the workplace. Some employers, meaning well, did everything they could to help.

> One employer moved an employee to three different buildings based on her complaints of mucus due to poor building air quality, then moved the entire organization to a fourth building. The woman's complaints continued unabated in the fourth building, and were continuing years after leaving the employer. Review of medical records indicated lifelong "mucus complaints," even when an infant. Her pediatrician wrote in his records, "This baby sure has a lot of mucus!"

Some persons moved from one state to another hoping to escape external factors such as pollen-provoking allergies. In all cases, however, after a grace period in some, mucus, sinus or hay fever complaints returned.

A nurse moved from New Jersey to the Sacramento area to escape her allergies. Unfortunately, it was not a choice relocation! Predictably, symptoms returned within a year, so she stopped working, blaming it on a wax remover once used on the floors at work by a janitor.

A clerk moved from a heavily wooded coastal area to a more southern, less heavily wooded coastal area where she resided and worked close to the beach. Her allergy symptoms had a grace period of two years, but by the third year had retuned to the level preceding the move. She blamed her allergy symptoms on something emanating from a photocopy machine.

In some cases, the degree of disability claimed was striking; persons were incapacitated by post-nasal drips. Persons who had disability insurance tended to stay home "on disability"; some individuals had not worked for years. Generally, the duration of the disability lasted as long as the policy coverage, six months, one or two years, or indefinitely. Several alleged to have remained in bed while disabled; some made it to the couch. Others could not work, but wrote novels, hiked, danced, and studied real estate or martial arts.

IV. VIRAL,
CARDIOVASCULAR, RESPIRATORY

The next-most-common groups, in more or less equal numbers, consisted of ordinary conditions that people have. Both doctors and patients seemed to forget about common illnesses at the mention of "work" or the word "toxics." Uncommon viral diseases were well-represented and included conditions, such as Meniere's Syndrome, shingles, hepatitis, AIDS, Guillain-Barre Syndrome, and post-polio syndrome. Cardiovascular diseases such as coronary artery heart disease and hypertension (high blood pressure) were noted, as were common respiratory conditions such as bronchitis, bacterial sinusitis, coughs, colds, sore throat, and flu.

In many cases, because of the emphasis on toxic injuries, the correct diagnosis was not made.

A 36-year-old auto body spray painter had to stop work due to increasing stumbling, weakness, and respiratory distress. He had polio as an infant, as well as several ankle surgeries, but he was a dedicated breadwinner and, as an adult, had rehabilitated himself and did heavy physical labor. He tired easily, and on physical exam, he was a tearful, anxious man with diffuse wheezing, muscular weakness, and atrophy. The patient, his wife and three small children were extremely concerned.

Unfortunately, this man was being sent from one physician and referral source to another, all of them focusing on some nebulous toxic constituent of auto body spray paint. However, he used full respiratory protection and worked in a state-of-the-art spray paint booth. Unfortunately, the only diagnosis which made sense in this patient was that of post-polio syndrome. Post-polio syndrome is a heartbreaking, profoundly debilitating condition that occurs decades after the original polio infection. Its cause is not understood, however, it is not caused by work.

V. SEXUAL PROBLEMS

In the sexual category, some of the complaints were sad; others had a humorous side. Women who presented with nausea and vomiting of pregnancy, "morning sickness," invariably were misdiagnosed if they reported to their doctors that they had a toxic exposure. This approach was dangerous because tranquilizers and headache medication were prescribed. There were miscarriages, ectopic pregnancies, cases of chlamydia, and side-effects of sex-change operations, all supposedly due to their jobs.

An unrecognized mini-epidemic of pubic lice was incorrectly attributed to toxics in the workplace. What the toxin was that victimized the pubic area during a building remodeling, no one could say.

A claim was submitted for "intoxication of a foreskin" so that expenses for adult circumcision could be paid by workers' compensation. Don't ask.

VI. NERVOUS SYSTEM DISORDERS

Diseases of the nervous system were as common as sexual disorders. Conditions such as migraine headaches, cerebrovascular accidents (CVAs), Bell's Palsy, petit mal epilepsy, and Parkinson's Disease were all mistakenly attributed to work.

A particularly sad case was that of a 65-year-old man who was, for 29 years, a manager or an insurance adjuster in the auto body repair industry. He noted the onset of fatigue, stiff joints, and a distinctly weakened grip, progressing to generalized weakness and an intention tremor. He could not hold a fork to his mouth without his hand shaking. He was desirous of receiving funds from workers' compensation, as he needed more of a pension than Social Security would provide.

When examined, there was good news and bad news. The bad news was that he had Parkinson's Disease. The good news was that he had never been treated, and many treatments likely to improve his condition were available.

He declined referral for treatment upon the advice of his lawyer.

VII. ASTHMA

Nine cases of asthma were considered to be unrelated to the workplace. In these patients, attacks occurred randomly and could occur at or during work, but the work and asthma were not related as cause and effect. The time course and occurrence of asthma were generally inconsistent with activities in the workplace; materials used on the job were generally not associated with asthma or used at times unrelated to asthmatic attacks. In general, physicians tended to remove this group from work too quickly, without really determining whether or not the asthma was related to work. As a result, all of them had continuing asthma but were without health insurance, jobs, or funds. Generally, they could have continued their work, some with little or no change in work practices, others with minor modifications that the employers probably would have accepted. Some persons hoped to receive lifetime medical benefits for previously existing asthmatic conditions, but just the opposite result occurred.

A man had sinus surgery and asthma in Europe. Two years after emigrating to this country, his asthma recurred. It was blamed on solvents that he sometimes used for cleaning parts, even though vapors from solvents were exhausted to the outside and his work area was well-ventilated. Two years after being removed from this job, his asthma was worse. He was labelled an asthmatic and could not obtain health insurance. He missed his job, and no one else would hire him.

VIII. OTHER CONDITIONS

Other cases that represented the spectrum of illness observed in an office practice included dermal reactions, such as vitiligo, irritation from dentures, contact dermatitis, athlete's foot, acne rosacea, and boot blisters. Six persons suffered from connective tissue disorders such as degenerative arthritis, rheumatoid arthritis, sarcoid, lupus, and monoclonal gammopathy together with peripheral neuropathy.

Small numbers of cases were present in other groups. Infections such as epididymitis, coccidioidomycosis and tuberculosis, and metabolic and endocrine conditions such as diabetes, porphyria, thyroid nodules, and sleep apnea presented frequently with the correct diagnosis, but no connection could be found between these conditions and the patients' jobs.

There were persons with asymptomatic microscopic hematuria who were given work-related diagnoses, as well as several cases of end-stage renal failure undergoing kidney transplants, requesting workers' compensation coverage for new kidneys.

Other cases consisted of elephantiasis, neurofibromatosis, ulcers, iron deficiency anemia, psychotic reactions, and a variation of Baron Munchausen's Syndrome. In Baron Munchausen's Syndrome, the patient simulates illness, often going from one doctor and hospital to another. There is a psychological craving for medical attention.

In our variant, a patient with a genetic disease (glucose 6-phosphate dehydrogenase deficiency) was told to avoid certain foods and drugs because life-threatening destruction of red blood cells (hemolysis) could be triggered. However, several months after being evaluated, this person successfully committed suicide by overdosing with drugs that triggered a hemolytic crisis.

While we did not observe any work-related toxic injuries in this group, it did provide a fascinating spectrum of medical patients.

IX. TRENDS AND CHARACTERISTICS OF NON-WORK-RELATED AND SPURIOUS CLAIMS

Claims were motivated by concern, alarm, drugs, and misdiagnosis. There were some disingenuous motives, as well.

Revenge

Some persons were terminated and wanted "revenge," presumably submitting a claim to annoy an employer and report malfeasances. Others were upset by mean-spirited relations between employers or coworkers and submitted claims not knowing what else to do.

Toxic Injury as an "Add-on" Diagnosis to Other Claims

In some cases, where claims were made for other injuries, such as musculoskeletal injuries or psychiatric stress, there appeared to be "add-on" claims for chemical exposure or toxic reactions. This was used like an option for extra insurance coverage when renting a car. The basis for such claims was that "something" was in the workplace, not that an exposure or injury had occurred. Claims were based on a list of potentially harmful substances found in the workplace.

Pension Request

Claim submission sometimes appeared to be a form of pension request. In some instances, the person had relocated to another state at the time the claim was submitted, precluding a return to the job.

Hysterical Reactions

Indoor odors, poorly ventilated buildings, and the use of bug sprays often triggered hysterical reactions in offices and warehouses (see Chapter 16).

"Allergy to Buildings or Environments"

A number of claims were submitted by patients who had already undergone "detoxification treatments" or other unconventional treatments costing several thousand dollars, hoping to have these bills paid out of workers' compensation funds. Typically, patients who were office workers were told that they were allergic to their buildings. Others were told of sensitivity to their work, such as "fumes" from typewriter correction fluid or "fumes" emanating from felt-tip pens. Persons often sought the help of doctors and other practitioners because of an "odor problem" which occurred following the application of a pesticide, painting, new furnishings, or installation of new carpeting. Some sought help upon noticing mucus from their noses.

Costly "detoxification" programs included megavitamins, saunas, baths, laxatives, enemas, diets, exercises, and herbs. Some were treated with "activated oxygen," germanium, or thyroid extracts. Others took sublingual and maintenance "antigens" such as "diesel exhaust" and "newsprint," following which many persons complained of being more, not less, sensitive to their environments. Typically they could no longer tolerate perfume, newsprint, supermarket aisles where detergents or toilet tissue were shelved, restaurants, dry-cleaners, or indoor building environments. Some developed hemorrhoids or dandruff. Several persons found comfort only on a beach and, though they desired vocational retraining, were not exactly sure how they could support themselves on the beach.

Persons who worked in industrial settings were told that they had developed allergies to "industrial chemicals" and could no longer work in their professions. This group frequently reported symptoms such as memory loss, confusion, feeling "spacey," and inability to concentrate. Several of them were diagnosed by a nonmedical doctor's "liquid crystal machine," computers, microscopes, and other bizarre instrumentation.

In some cases this quackery was unrecognized, and persons received disability payments, often substantial. Such persons were "diagnosed" as having "environmental illness or allergy" and believed themselves to be "permanently sensitized" to the "environment." Buzz words frequently appearing in these claims were "immune system damage," "immune disregulation," "psychic stress" as the result of exposure to a molecule of something, brain damage, chronic headaches, and lost intellect. The claims of loss of intellect typically involved a janitor's assistant who had lost the intellect of a physics professor.

Strange Symptoms and Rat/Mouse Disorders

Compared to the work-related cases, non-work-related symptoms involved several organ systems instead of one. Symptoms favored the bizarre and unusual.

"Chemical allergy" symptoms noted above (memory loss, confusion, "spacey" feeling, inability to concentrate, and immune disfunction) were reported frequently in the non-work-related cases and rarely in the work-related ones.

Symptoms of toxic reactions more often than not matched those described as occurring in rats and mice on Material Safety Data Sheet forms or other toxic information literature. In particular, all symptoms that were ever reported in human or animal testing without regard to dose or route of exposure were described by these persons. While the symptoms were true-to-form for overdosed rodents, they were not convincing or realistic descriptions of workplace exposure symptoms.

Inconsistent Time Course

Common to these cases, symptoms worsened *after* removal from exposure. New symptoms developed, often long after the exposure had ceased. Sometimes symptoms did not develop until after legal advice.

A worker briefly entered a room where a small ethylene oxide leak had been identified and repaired a few days before. After leaving the area, she complained of headache and eye irritation. Levels of ethylene oxide were measured within minutes of her departure, and the substance was undetectable. A month after the incident, she complained of sensitivity to tobacco smoke and auto exhaust and was unable to work.

Drug Effects Common

Drug symptoms were more common in the non-work-related group. In these cases, workers were more likely to be on medication with side-effects compatible with their symptoms and overexposure to prescription, over-the-counter drugs, or illicit drugs were often not recognized. Drug use accounted for most symptoms attributed to "toxic" exposure.

Misdiagnosed Medical Problems

These were workers who had legitimate medical problems but had been incorrectly diagnosed. Examples of such conditions erroneously diagnosed as work-related were tinea pedis (Athlete's foot) with severe -id reaction on the hands, AIDS, influenza, and pregnancy.

Prominent Psychological Problems

Significant non-work-related psychological problems motivated the complaints in this group.

> A worker complained of skin lesions that he said began after he started working. Examination revealed small, circular scars. Review of medical records indicated that similar lesions, which looked like cigarette-burns and occurred only in easily-accessible areas, had been identified as fresh cigarette burns by a physician at a prior job.

It was the patient's intent to simulate the remains of a bullous dermatitis (blistering skin reaction) from a fumigant.

23

Management of On-the-Job Injuries. Getting Back to Work – the State Fund Mutual Way

Andrew C. Meuwissen and Jacqueline K. Ross

State Fund Mutual, Eden Prairie, MN

KEY TOPICS

I. HISTORY OF STATE FUND MUTUAL

Minnesota's State Fund Mutual was created in conjunction with the state's Workers' Compensation Reform Act of 1983. This act dramatically changed both the statutory benefits and how they were to be administered.

This reform intended to address a progression of problems, including increased litigation costs, benefit levels increased by the legislature, comparatively higher medical cost in Minnesota, and adoption of an expensive annual cost-of-living adjustment that made increasing costs part of the permanent structure. The combined results of these forces were devastating financially to Minnesota employers.

Private insurers complained that investment returns were too low, rates were inadequate, and benefits were far too high. The Minnesota legislature decided to provide an insurance choice to employers in Minnesota: an independent voice in the workers' compensation system, a true specialist carrier providing workers' compensation coverage. Thus the idea of the State Fund Mutual was born in Minnesota, and it was formed in September 1983.

The company was incorporated as a mutual insurance company, with every employer who purchases insurance becoming a voting member of the corporation. The State Fund Mutual wrote its first policy effective April 1, 1984. It has grown to be the second largest writer of workers' compensation coverage in the state.

The purpose of the State Fund Mutual was to foster increased competition in the insurance industry marketplace, thereby stabilizing rates to employers. By providing an additional market and specializing in workers' compensation, the company could keep rates competitive and focus efforts on effective workers' compensation cost control. The State Fund's mission: working cooperatively to create the preeminent workers' compensation delivery system for Minnesota, distinguished by quality, cost-effective services — preserving human and economic resources through innovative professional people.

The State Fund is a highly specialized insurance company, operating only in Minnesota and offering only workers' compensation services. Specialization was a smart strategy because of Minnesota's unusually complex workers' compensation administration system. A cost-effective operation within this complexity requires significant premium volume and market share.

This specialization provides a unique marketing advantage. By focusing on one line of insurance, State Fund Mutual did not need to juggle competing market and management strategies. Personnel could be trained to know Minnesota's benefit system well. Management decisions, short- and long-term, could be focused on making this system work well in Minnesota, without ever-changing (and frequently out-of-state) "home office" decisions about local administration of a little-understood statutory scheme. Let's examine some statistics.

Size of Company

Employees (as of July 1990)	124
Policies	5400 employers
Volume (as of July 1990)	Average 215 per week, First Reports of Injury
	30% result in lost-time benefits, 70% for medical treatment only
	4800 open claims

Financial Success of State Fund Mutual, 1989:

Controlled Growth	Net premium writings up 9.7%
	Remained second largest writer of workers' compensation in the state
Combined Ratio (Cost of claims and expenses relative to premium income)	90.3% (68.3 loss ratio; 22% expense ratio). Compare to industry averages of 113% in Minnesota and 120% nationally
Net Income	Increased 93% over 1988 to $4.7 million
Policyholder surplus	Increased 47% over 1988
Dividends Paid to Policyholders	Increased 55% over 1988 to $1.4 million

The State Fund Mutual strongly believes in the concept that workers' compensation insurance coverage is really a benefit delivery system. The welfare of the worker is a most important objective. Servicing the worker's needs is not inconsistent with the best interests of the employer. For example, providing benefits on a timely and consistent basis is a benchmark.

A close working relationship with policyholders is a hallmark of State Fund's success. Smaller employers, particularly, need education and closer attention. Of State Fund's policyholders, 78% have premiums of less than $10,000.00; 50% have $2,500 or less in premiums. The State Fund is able to provide reasonable premium rates because of the intense interaction between the company and the policyholders in both loss prevention and loss control.

Premium Size of Policy Holders, 1989:

More than 34% $1,000 or less in premiums
50% $2,500 or less in premiums
78% $10,000 or less in premiums

Policies Written by Business Classification, 1989:

26%   Government Organizations
21%   Retail/Wholesale Trade
16%   Services
14%   Agriculture

10% Manufacturing
9% Construction
4% Transportation

II. A COMMITMENT—EARLY RETURN
TO WORK

The State Fund's philosophy centers primarily around one very important concept: early return to work. It is a well-known fact that the longer an injured person remains off work, the harder it is to return that employee to productive activity. This "delayed recovery syndrome" may be the product of physical deconditioning, or an employee's feelings of rejection, performance anxiety, or comfort in a nonworking lifestyle. For whatever reason, a delayed return to work is very expensive.

Benefit of Early Return to Work—Case
Example: Harry

An example of the cost differential between maintaining a successful return to work program versus none at all will become apparent after reviewing the following actual State Fund Mutual case example.

Harry, a 56-year-old truck driver, injured his back when his truck hit a bump in the road. If the employer had had a successful return to work program, Harry might have been able to return to work in as little time as two to four weeks in a limited capacity as set forth by his physician. Without a successful return to work program, Harry may not have been able to return to his regular job until after a much longer healing period, thereby costing the employer significant dollars. Harry was diagnosed as having two herniated discs. The employer indicated his willingness to take Harry back to work, even with restrictions, as soon as the physician released him to light duty. Surgery was not recommended. On that basis, the claim reserves were set as follows:

Indemnity benefits	$15,200
Medical benefits	2,750
TOTAL	$17,950

Harry returned to work shortly after the accident occurred. However, he was again disabled several months later due to a flare-up of his condition. After this flare-up, the employer refused to take him back into a modified job. Harry's age, education level, and limited skills meant that successful job placement would be difficult. Meanwhile, Harry's medical treatment costs were escalating. The claim reserves were revised to:

Indemnity benefits	$ 99,415
Medical benefits	22,910
TOTAL	$122,466

The employer was advised of the additional costs of this claim due to the company's not providing Harry with a job. A permanent job was then offered to Harry which included job modification: in this case, a special seat. As a result of the State Fund working with this company to develop light and modified job tasks, the expected payout was revised to $29,684.

Policy Application Review

The State Fund has incorporated the concept of return to work in all of its operations, starting at the policy application level. Each applicant must commit in writing to participate in a viable return to work program. This concept is further enforced by a specific policy endorsement that confirms the obligation of the employer to participate in such a program.

The Underwriting Department reviews the application for insurance with an eye toward active participation and management on the part of the potential insured of their safety and return to work programs. If the underwriter cannot determine the extent to which such programs exist, a representative of the Loss Prevention Department makes a personal visit to the potential insured and assists in audits of the premises for hazards and reviews all existing risk management programs. Loss Prevention takes an active role in assisting the insured with their risk management programs, including return to work and second injury fund registration. Loss Prevention operates under "Total Loss Control" when screening potential insureds and servicing existing insureds to help visualize how losses might occur. It provides tools to understand how losses may escalate and provides a forum to explain loss concepts and principles to the insured. The value of prescreening is identifying the degree of loss potential and unexpected risks and determining the likelihood of account cooperation and the degree of commitment to State Fund's program.

Once it is determined that the potential insured meets the State Fund Mutual's underwriting requirements, a policy is issued. When servicing existing accounts, Loss Prevention stays in contact with the insured to help implement new programs and recommendations to minimize loss and to protect the State Fund's interests.

Employer's Return to Work Program

A return to work program is a system developed by the employer and representatives from State Fund's Loss Prevention Department to return the injured worker, within medical restrictions, to the workplace as soon as the employee is capable of performing some work activities for the employer. The program is designed for cost-effective and successful management of employees injured on the job.

> Workers' compensation is a big expense. That's why we've made it as much a part of our business plan as any other aspect of the company. Our return to work program is a major part of that plan. When one of my employees gets hurt, I want him or her back quickly. These people take great pride in their work and are a tremendous asset to me. I need them, and they know that. They make this place tick. They can't help out when they're at home.
>
> [Plus,] I just received our dividend check. That's great incentive.
>
> Doug Kracht, Vice President and General
> Manager of the Hubbell House, a Restaurant

State Fund recommends that the employer's return to work program include:

1. A statement expressing management's support for the program.
2. The purposes and advantages of using a Return to Work Program.
3. The conditions under which it is appropriate to use the Return to Work Program.
4. Procedures for providing proper and immediate medical attention to injured employees.
5. Step-by-step instructions explaining what to do and whom to notify in case of work injury.
6. Responsibilities of the injured employee, co-employees, supervisors, claims coordinator, and management in the return to work process.
7. Light-duty job identification.
8. The proper flow of paper work and communications.

This written program may also include the employer's expectations of its primary medical provider, qualified rehabilitation consultants, and the insurance company.

Primary provider. A primary provider is a physician or clinic selected by the employer to provide medical care to their injured employees. In Minnesota, an employee is free to choose a medical provider but frequently looks for recommendations from the employer. An employer who has established a relationship with a reputable provider in advance of a work injury can offer an employee prompt, quality medical attention. By establishing a relationship with a particular medical provider, the provider is familiarized with the nature of the business and its jobs and whom to contact with status reports. The parties work together, outline expectations, and make certain that the employer's individual and specific needs for quality medical care and return to work are being met. Open communication is encouraged between the provider, the employee, the insurance company and the employer, resulting in a more successful return to work program. The Underwriting, Loss Prevention, and Claims Departments can be of great help in selecting a local provider to offer medical services.

In identifying a primary provider, the following are considered:

1. The physician's ability to deal with occupational health problems.
2. The physician's professional reputation and standing.
3. Whether the physician has a fundamental knowledge of the workers' compensation system and is willing to work within it.
4. The physician's willingness to get to know the employer's facilities and available work.
5. Convenience of the physician's hours and the physician's willingness to see the employees as soon as possible, with little waiting time in the physician's office.
6. The physician's willingness to communicate with other physicians and the insurance company about medical progress and physical capabilities.
7. The physician's understanding of the employer's return to work program.
8. The capability of the physician to provide pre-placement physicals, Second Injury Fund registration, audiograms, and other loss prevention needs of the employer.
9. The physician's willingness to meet with the employer on a regular basis to determine if problems are arising and if adjustments to the program need to be made.

Once the employer has chosen a primary provider, that provider should visit the facilities as often as necessary to become familiar with the operations of each department where injuries are likely to occur. The physician can help identify light-duty jobs that injured employees could return to without causing further injury to themselves or being a threat to other safety. Thinking about this in advance of injury is key.

Employer's claims coordinator. A claims coordinator should be selected by the employer to run the internal return to work program and be the main contact with the insurance company's representatives. With one person responsible for this aspect of the claims management program, the program becomes more organized, runs smoothly, and is used consistently. A claims coordinator should be familiar with the proper accident reporting requirements, the workers' compensation system, company policies, the specific jobs and work processes involved, and the primary provider. This individual should also possess strong verbal and written skills, project a positive attitude toward the employer's program, and deal with the injured employees sensitively and consistently. Most critical, however, is that this individual must have top management support and backing.

Identifying light-duty jobs. The identification of light-duty jobs by the employer is a critical step in the return to work program. These types of jobs provide employees with an opportunity to begin working again as soon as possible within medical restrictions. The employee benefits by returning to a productive role in the work force, and the employer benefits by lowering claim costs as well as maintaining productivity.

Light-duty jobs are best identified by using a team approach. The claims coordinator, the primary provider, and the loss prevention representative from the insurance company can provide extensive insight into formulating light-duty jobs. In many instances, an employee can return to his old job with slight modifications. In other instances, the employee's skills may be utilized in a number of different positions within the employee's medical restrictions.

Prompt medical care. Once an injury occurs, prompt medical care by the primary provider begins and can make the difference between a successful recovery period and one fraught with problems. The injury must be reported immediately to the Claims Department to allow the claims representative to begin investigation. The claims representative will then contact the insured's claims coordinator to get the details of the accident. Thereafter, the injured employee and the treating physician are contacted.

If the medical condition is of an unusual nature, the claims representative will meet with one of State Fund's on-staff physicians or chiropractors. The on-staff physician or the claims representative will contact the treating physician with appropriate questions concerning the nature of the injury and advise the treating physician of the insured's light-duty job opportunities. This has proved to reduce dramatically the length of time an injured employee is off the job.

Lost-Time Injuries, 1989:

25%   Low back
19%   Wrist, hand, or fingers
Note: The majority of these injuries occur in construction and other
   "heavy industrial" settings.

Continual contact. If the injured worker cannot immediately return to
work in his or her regular capacity, the claims representative maintains
continual contact with the insured, the injured worker, and the treating
physician. As the recovery of the injured worker progresses, the claims
representative and an on-staff physician continually monitor the ability of
the worker to return to some sort of employment. Once any type of
restriction is issued, the claims representative, the loss prevention representative, a qualified rehabilitation consultant (if necessary) and the insured work together in successfully returning the injured worker to productive employment within the employer's operation. If an employer cannot
take the injured employee back to work in any capacity, outside placement
must begin.

   This type of diligent work—also described in the case example of
Harry—by the State Fund with its policyholders vividly demonstrates the
reduced costs that can result from returning injured workers to light-duty
jobs at their employers. The State Fund's belief that a successful return to
work program reduces the average costs of claims is supported by independent research done by Dr. Alan Krueger of Princeton University. He found
that employees whose employers were insured by the State Fund Mutual
returned to work 15% sooner than average. The State Fund Mutual's
success in handling these injuries means shorter duration of disabilities,
lower costs and higher productivity to employers.

Benefits to Employees from Early Return to Work

In most cases, the incentive to return to work revolves around benefits paid
by the employer. Most employers pay benefits such as health insurance,
pension and profit sharing and the like for a limited period of time if an
employee is injured. Should an employee's disability reach a certain pre-
determined length of time, such benefits may be in jeopardy.

   In most cases, the employee's compensation rate is very similar to their
after-tax take-home pay. Only for highly paid employees does an inequity
occur. Highly paid workers realize financial benefits upon return to work
because there may be a difference between the compensation rate while
disabled and their full wages when back on the job.

The only additional financial incentive an injured employee may have in returning to work occurs when he or she suffers a permanent impairment as a result of the work-related injury. A lump sum benefit may be paid 30 days after a successful return to work.

The real incentive for an employee to return to work is social and psychological rather than financial. The quicker an injured employee is returned safely back into the working environment, the more productive and accepted he or she feels. Their own desire to recover and return to productive employment only hastens the rehabilitation process, allowing them to become productive members of society again.

After my injury, I found State Fund Mutual's service to be tremendously prompt. They started working on my behalf immediately, and continually updated me on the process, answering my questions in detail. Because of their concern, my injury disrupted my life as little as possible.

Archie Rice, Jennings Red Coach Inn, Inc.

What I liked best about the service I received from State Fund Mutual is that they showed a genuine concern for my well-being. They didn't just treat me like "another claimant." They were concerned, provided timely service, and kept me well-informed about the entire process.

Jean Oldakowski, Thrift-Way, Inc.

I love my work. It's fulfilling to me. I don't know what I would have done if the Department hadn't taken me back in the field I enjoy so much.

Gary Scofield, Deputy Sheriff

III. EXTERNAL AND INTERNAL COOPERATION

What further makes the State Fund Mutual successful as a worker's compensation insurance company, and specifically in its return to work and total claims management, is the complete cooperation between employer, agent, and State Fund Mutual employee. Along with the external cooperation and educational activity, there is also the internal, departmental company cooperation in the State Fund Mutual. We have already mentioned the work of the underwriter and loss representative, but all departments are involved in the selling and implementation of the total claim management and return to work process. Field payroll auditors visit the employers for payroll auditing and proper hazards classifications. They also review and verify return to work philosophy and programs.

In-House Medical and Legal Resources

The State Fund Mutual utilizes unique medical case management and legal in-house resources. From day one of each disability case, the State Fund Mutual insists that the injured employee receive quality and appropriate medical care. The State Fund Mutual provides extensive utilization and cost review of medical treatment as well as medical case management.* State Fund Mutual's trained professionals examine each medical charge for:

- Compensability
- Prevailing and appropriate billing practices
- Usual and customary charges
- Reasonable and necessary services
- Overutilization
- Medical Fee Schedule maximums

Claims representatives confer with an in-house physician or chiropractor and a staff attorney to discuss medical and legal strategies for return to work, quick and fair settlements, avoiding litigation, etc. The interaction of company personnel is done daily, face to face, allowing cases to progress faster with time delays eliminated. The company reacts quickly to changes in situations. Proactive strategic planning is utilized immediately and is ongoing to resolve issues and establish a great trust level.

Proactive Service Pays

Efficiency— Results of Studies

Timely Benefits Ratio (Speed of response to claims by either paying or denying benefits)	Faster than the average insurance company in Minnesota
Rehabilitation Costs	60–65% of average in Minnesota
Gross Average Claim Cost	Substantially less than "State Trended Gross Average Claims" cost for mature years as well as recent accident claims
Legal Cost (no studies conducted)	Believed to be lower than those of competition
Cost Management Program*	Savings of $1.2 million in medical provider costs
Medical Case Management*	Savings of $800,000

*Total reduction of all medical services billed
17% for $2 million savings; increase of 6% over 1988

Return to work activity is at the heart of the Minnesota Workers' Compensation legislation and the State Fund Mutual's practices. The State Fund Mutual's success, financially and operationally, can be attributed to its return to work efforts.

24

Getting Workers' Compensation Back to Work

KEY TOPICS

I. Non-Work-Related Claims are Increasing
   - Health Claims Masquerade as Workers' Compensation Claims
   - Drug Problems are Increasing
II. Unworthy Toxics and Drug Claims Distract from Legitimate Ones
III. More Serious Injuries are Anticipated
   - Off-Shoot of a Sluggish Economy
   - Drug-Related Accidents
   - AIDS
IV. A More Responsive and Preventative System is Needed
V.   Workers Are People, Too

While the next decade will be challenging for business, it will try the limit of a workers' compensation system that differs from state to state, and in the private and public sectors. Some carriers will cope better than others, but all can expect some serious problems.

I. NON-WORK-RELATED CLAIMS ARE INCREASING

Health Claims Masquerade as Workers' Compensation Claims

People have aches and pains, colds and flu and medication side-effects. These conditions, not due to workplace toxic exposures, are commonly submitted as work-related claims. What can be done about them? The industry needs to weigh the pros and cons of combining work-related insurance with health insurance. Unfortunately, major drawbacks that I have observed are that needless treatments and over-medication increase in proportion to insurance coverage. If medical and workers' compensation are combined, I would recommend against prescription coverage unless strict criteria are used and enforced. If health and compensation insurance are separate, objective diagnostic criteria are needed to eliminate medical claims inappropriately submitted as work-related claims.

Drug Problems Are Increasing

Drug problems frequently present as work-related claims with drug side-effects for both prescription and illicit drugs incorrectly attributed to chemicals at work. Symptoms of drowsiness, weakness, malaise, dizziness, heightened anxiety, nausea, loss of concentration and memory loss can all result from medication. Common examples are pain medicines, tranquilizers, anti-anxiety drugs or drugs for sleep. Drug addiction or withdrawal poses other problems.

II. UNWORTHY TOXICS AND DRUG CLAIMS DISTRACT FROM LEGITIMATE ONES

Toxics claims, presently an umbrella for all sorts of weird claims, are likely to increase as lay-offs increase. The claims examiner needs to be wary of those potentially very expensive toxics claims, often involving groups, where lifetime disablement is alleged from a trivial or non-event. In contrast, for genuine cases of exposure, there is no cogent mechanism for rapid evaluation and reassurance. Most of these patients get lost in the system. Undoubtedly, the majority of work-related claims will continue to be orthopaedic, but medication problems in these patients are on the rise. An injury may be trivial, but drug side-effects, over-treatment and addiction are not, and prevent the return of a useful worker.

III. MORE SERIOUS INJURIES ARE ANTICIPATED

Off-Shoot of a Sluggish Economy

A sluggish economy, already upon us, is heralded by maintenance cutbacks, outmoded equipment and a bare-bones work force, conditions likely to result in major accidents, serious injuries and death claims. The insurer can avert tragedies by demanding evidence of good work practices and well-maintained facilities.

Drug-Related Accidents

Overuse of both prescription and illicit drugs are related to both vehicular and non-vehicular on-the-job accidents, often fatal. People experiencing drug-related symptoms are a danger to themselves and to other workers. Policies that ensure clear and safe thinking on the job could reduce this risk.

AIDS

Claims for treatment in deaths from Acquired Immune Deficiency Syndrome (AIDS) will increase in two ways. Claims in health care workers, law enforcement and prison workers may increase. AIDS has also been acquired during treatment for on-the-job injuries, such as from blood transfusions during surgery. Although the supply of blood for transfusion may now be safer than previously, autologous transfusions should be encouraged, and needless surgery eliminated.

IV. A MORE RESPONSIVE AND PREVENTATIVE SYSTEM IS NEEDED

Not all worker compensation programs enjoy the success described in the previous chapter. Many workers' compensation programs are unwieldy and bureaucratic and no one is well served. Resources need to be redirected to the injured worker and to getting him back to work.

Few workers' compensation systems incorporate prevention. The experience of Minnesota State Fund Mutual shows not only that prevention does pay but that systems that are proactive and prevention-oriented can be especially profitable for the insurer, the worker and the employer. Everybody wins!

V. WORKERS ARE PEOPLE, TOO

Workers are *people,* too. The ones I meet are lost in a system that can be distant and impersonal. The Director of Human Resources, Medical and Benefits of a large corporation remarked about imprinted hats that the workers received at a company function. "Although these hats are nice", he said, "what our workers really need are hugs." I wonder how many compensation claims would disappear if employers and employees were nicer to each other. Since happy people take fewer drugs, are better workers and make fewer mistakes, one exposure all would benefit from is a big dose of human kindness.

Appendix

Trace Metals, Organics, Solvents and Biological Indicators in Body Fluids

Test	Usual Sample	Sample Volume	Sample Container	Special Notes
Acetone	Blood	2 mls	L	
Alcohols: Acetone, Ethanol, Formic Acid, Isopropanol, Methanol	Blood	4 mls	L	
Aluminum	Serum	2 mls	SST	
	Urine	50 mls	B	
Antimony	Urine	50 mls	B	
Aromatic Solvents: Benzene, Ethylbenzene, Toluene, Xylenes, Styrene	Blood	4 mls	L	1,3,4
Aromatic Solvent Metabolites: Free & Conjugated Phenol, Hippuric Acid, o-Cresol, Methyl Hippuric Acid, Ethylphenol, Phenylglyoxylic Acids	Urine	50 mls	C	1,3
Arsenic	Blood	2 mls	L	
	Urine	50 mls	B	
Barium	Urine	50 mls	B	
Benzene	Blood	2 mls	L	4
Beryllium	Urine	50 mls	B	
beta-2 Microglobulin	Serum	2 mls	SST	
	Urine	50 mls	A	2
Bismuth	Blood	2 mls	L	
	Urine	50 mls	B	
Cadmium	Blood	2 mls	L	
	Urine	50 mls	B	
Carboxyhemoglobin	Blood	2 mls	L	
Carbamate Pesticides: Carbaryl, Carbofuran, Landrin®, Methomyl®, Propoxur	Serum	4 mls	SST	3
Chlorinated Pesticides: alpha Chlordane, gamma Chlordane, Heptachlor, Heptachlor Epoxide, Oxychlordane	Serum	4 mls	SST	3

(Continued)

219

Test	Usual Sample	Sample Volume	Sample Container	Special Notes
Chlorinated Solvents: Perchloroethylene, Trichloroethane 1,1,1, Trichloroethylene	Blood	4 mls	L	3,4
Chlorinated Solvent Metabolites: Trichloroacetic Acid, Trichloroethanol	Urine	50 mls	C	1
Cholinesterase: RBC	Blood	2 mls	SST and L	
Serum	Serum	2 mls	SST	
Chromium	Blood	2 mls	L	
	Urine	50 mls	B	
Cobalt	Blood	2 mls	L	
	Urine	50 mls	B	
Copper	Serum	2 mls	SST	
Creatinine	Serum	2 mls	SST	
	Urine	50 mls	B	
Cyanide	Blood	2 mls	L	
	Urine	50 mls	B	
Ethylbenzene	Blood	2 mls	L	4
Ethyl Ketone	Blood	2 mls	L	
Fluoride	Serum	2 mls	SST	
	Urine	50 mls	B	
Formic Acid	Urine	50 mls	C	
Hippuric and Methylhippuric Acids	Urine	50 mls	C	1
Hydrocarbon Solvents: Aliphatic (C_4-C_{10}) and Aromatic	Blood	4 mls	L	3,4
Iron	Serum	2 mls	SST	
Isopropanol	Blood	2 mls	L	
Ketones: Acetone, Methyl Ethyl Ketone, Methyl-n-Butyl Ketone, Methyl-iso-Butyl Ketone, Cyclohexanone, Mesityloxide, Diacetone Alcohol, Isophorone	Blood	4 mls	L	3
Lead	Blood	2 mls	L	
Magnesium	Serum	2 mls	SST	
Manganese	Blood	2 mls	L	
	Urine	50 mls	B	
Mercury	Serum	2 mls	SST	
	Urine	50 mls	B	
Methanol	Blood	2 mls	L	4
	Urine	50 mls	C	
Methemoglobin	Blood	2 mls	L	
Methyl-iso-Butyl Ketone	Blood	2 mls	L	
Molybdenum	Urine	50 mls	B	
Nickel	Urine	50 mls	B	
Organochlorine Pesticides: alpha Chlordane, gamma Chlordane, Heptachlor, Heptachlor Epoxide, Oxychlordane	Serum	4 mls	SST	3

Test	Usual Sample	Sample Volume	Sample Container	Special Notes
Organophosphate Pesticides: Diazinon, Parathion, P-Nitrophenol, Dichloryos, Methylparathion, Chlorpyrifos, Guthion®, Phorate, Triclofos	Serum	4 mls	SST	3
PCB's: Polychlorinated Biphenyls-Arochlors: 1016, 1221, 1232, 1242, 1248, 1254, 1260	Serum	4 mls	SST → P or SST	
Perchloroethylene	Blood	2 mls	L	
Phenol	Urine	50 mls	C	
Porphyrins: Coproporphyrin, Uroporphyrin	Urine	50 mls	B	
Retinol Binding Protein	Serum	4 mls	SST	
	Urine	50 mls	B	
Selenium	Blood	2 mls	L	
	Urine	50 mls	B	
Silver	Blood	2 mls	L	
Styrene	Blood	2 mls	L	4
Tellurium	Urine	50 mls	B	
Thallium	Blood	2 mls	L	
	Urine	50 mls	B	
Thiocyanate	Serum	4 mls	SST	
	Urine	50 mls	C	
Titanium	Urine	50 mls	B	
Toluene	Blood	2 mls	L	4
Trichloroethane	Blood	2 mls	L	4
Trichloroethylene	Blood	2 mls	L	4
Vanadium	Urine	50 mls	B	
Xylene	Blood	2 mls	L	4
Zinc	Serum	2 mls	SST	
ZPP (Zinc Protoporphyrins)	Blood	2 mls	L	

1. Refrigerate Specimen until shipped.
2. Special container with Tris buffer, pH must be above 6.0, contact laboratory.
3. Group test includes commonly seen compounds. Contact laboratory for specific details on other compounds.
4. Also consider urine metabolites for indications of chronic exposure.

Samples listed are preferred, other samples may also be tested. Contact client services at the laboratory for other acceptable samples and specimen collection containers.

Sample Containers:
A: 50 ml Urine Container with Tris buffer
B: 50 ml Urine Container with EDTA
C: 75 cc Urine Container-No Preservative
L: Lavender Top Tube with EDTA
P: Plastic Screw Top Tube
SST: Serum Separation Tube-10 ml vacutainer
SST → P: 10 ml vacutainer drawn → serum poured off
Information furnished by BioTrace Laboratories, Salt Lake City, Utah. 1-800-950-8722

Glossary

Absorption—whether the substance can enter the bloodstream through the skin, lungs or gut, etc.

Acids/Corrosives/Reactive Chemicals—substances intrinsically hazardous upon contact

Active—causes a health effect

Acute—sudden or short term

Affect—emotional state

AIDS—Acquired Immune Deficiency Syndrome, caused by viruses, lethal

Air Contaminants—anything present in air that is not carbon dioxide, oxygen, nitrogen or water. In this book referred to as chemical, but may be dust, fungi, bacteria

Alcohol—used to describe ethanol-containing beverages

Alcohol Intoxication—a legal definition regarding the amount of blood alcohol (blood alcohol concentration) that impairs function of the central nervous system

Aldehydes—type of chemical such as formaldehyde, glutaraldehyde

Allergic History—reactions provoked by foods, dusts, pollens, drugs, chemicals, or anything mediated by some aspect of the immune system

Allergic Rhinitis—symptoms of watery eyes, runny nose, cough, such as hay fever, is on an allergic basis to pollens, mites, dusts, not "toxic chemicals"

Allergy to Buildings or Environments—unconventional philosophy where diagnosis is made without objective medical or test evidence

American Conference of Governmental Industrial Hygienists (ACGIH)—non-governmental, respected standard bearers who make recommendations for safe workplace air levels (TWA TLVs)

Ammonia—NH_3—used as cleanser in household ammonia, as refrigerant and as ammonium salts in fire extinguishers and permanent wave solutions, for example.

Analysis—laboratory testing for one or more specific substances

Analyte—substance to be analyzed

222

Anemia—low red blood cell count, may be due to many causes, including lead poisoning

Anemia, Aplastic—see Aplastic Anemia

Angioneurotic Edema—allergic swelling of the throat, often accompanied by swelling of various portions of face and hives

Angiosarcomas—specific, unusual blood vessel tumors found in livers of vinyl chloride monomer workers in 1970's

Animals—guinea pig: furry animal with no neck, eats and poos; hamster: cute mouse-type animal with predilection for little ferris wheels; rats and mice: small rodents generally regarded as pests in their natural setting, cute and furry but not your typical worker. Special strains of laboratory animals are inbred for experiments.

Animal Experiments—overdosing animals especially bred for this purpose under defined conditions

Anoxic—lack of oxygen

Antagonism—counteraction

Anticholinesterase—a substance, such as certain organophosphate pesticides, which inhibits the enzyme, cholinesterase

Antidepressant—drug used to treat depression

Antidote—a substance that specifically counteracts or reverses the effect of another

Antihypertensive—drug used to lower elevated blood pressure

Aplastic Anemia—a serious condition in which the body ceases to manufacture blood cells and platelets

Apportionable—or apportionment, is the extent to which a disability or a level of impairment is due to different factors. These factors include disabilities that pre-existed the health condition presently under consideration such as the disabilities resulting from prior or pre-existing injuries, pre-existing medical conditions or contribution of personal lifestyle factors such as smoking or alcohol abuse. The degree of disability is divided or apportioned between these non work-related and work-related factors.

Archived—this term applies to specimens frozen and stored until needed

Arsenic—element, present in food in low amounts, toxic in high quantities

Arthritic—pertaining to arthritis, disorders of the joints

Asthenia—extreme fatigue or listlessness

Asthma—disease where the bronchi and bronchioles narrow, tendency may be inborn or acquired, hallmark on physical exam is wheezing

Asthma, Non-Work Related—caused by unknown factors; bacterial infections; allergy to pollens, foods, sulfites, etc.; medication/drug reactions

Asthma, Occupational or Industrial—provoked by workplace factors. Typical causes include exposure to: isocyanates and constituents of uncured plastics; sulfites as chemicals or off-gassing; natural or synthetic resins or gums; organic dusts.

Auscultated—listened to

Autologous—with respect to blood transfusion, person donates blood for their own use

Axillae—armpits

Background, Chemical—the amount of chemicals in air, breath, blood or otherwise in our bodies that is present without a particular event or exposure; the level below which 95 percent of the values from non-occupationally exposed individuals will fall; the general population range

Bacterial—pertaining to bacteria, which are living organisms

Baseline Specimens—specimens from a person, such as blood or urine, collected before starting a job and before a potential exposure. These can be stored (archived) frozen until needed.

Baseline Status—condition prior to a job, circumstance or an exposure; state of health prior to an illness

Baseline Test—measures status (such as airway function) before an exposure has taken place

Bell's Palsy—sudden one-sided paralysis of the face, presumed to be due to an immune or viral disease involving the seventh (facial) cranial nerve. Facial expression is distorted and recovery is variable.

Benzene—6-membered carbon ring (aromatic) compound formerly used as a solvent, but restricted because of its association with aplastic anemia. Naturally present in food and vegetation; and in low percentages, gasoline. Also present in cigarette smoke.

Benzodiazepines—a large class of "antianxiety" tranquilizing drugs. The oldest drugs in this class are Librium and Valium with many others presently in use. These drugs are subject to abuse with a propensity for addiction and symptoms of heightened anxiety. Sometimes seizures develop upon withdrawal. Some persons develop memory loss and other evidence of central nervous system impairment while on treatment; the elderly are especially susceptible.

Beta-blockers—drugs which block specific adrenergic receptors known as Beta receptors, used to lower blood pressure

Beta-2-microglobulin—a protein in the urine once thought to be useful in assessment of cadmium exposure but of questionable usefulness at present

Bioavailable—see Bioavailability

Bioavailability—denotes whether a drug or chemical can get into the body; denotes whether a drug or chemical can be active (or have an effect) once it enters the body

Biological Exposure Indices (BEIs)—BEIs are levels of chemicals and their metabolites in various human specimens which should not be exceeded in order to avoid adverse health effects. Guidelines for specific chemicals with their recommended collection timing are listed in Table 20-1.

Biological Half-Life—the amount of time it takes for half of the substance in the blood or body to disappear

Biologic Monitoring—tests that help to determine whether an exposure has occurred by examining samples such as blood and/or urine from the person and analyzing them for chemicals or chemical metabolites.

Biologic Reaction—any measurable response in a living thing

Blood Carbon Monoxide—the amount of carboxyhemoglobin in the blood

Blood Lead—amount of lead in blood measured from a blood sample

Body Burden—the amount of a drug or chemical present in the body

Bronchi—progressively narrower airways leading to the air sacs of the lungs

Bronchioles—the smallest airways that lead to the air sacs of the lung
Bronchitis—inflammation of the bronchi, usually a bacterial or viral infection, less commonly work-related
Bruits—noises in the blood vessels made by narrowing, obstructions or malformations
Bullous—large blisters
By-product—a chemical or chemical mixture generated as an off-shoot or result of a reaction intended to manufacture something else

California Labor Code 3212.1—regulation applies to cancer in firefighters—cancer is presumed to be work-related if worker has reasonably been exposed to a known carcinogen (IARC) that produces that kind of cancer
Cancer—malignant tumor of serious nature, often leading to death if untreated
Cancer Prevention Studies I and II—surveys conducted by American Cancer Society on smoking and mortality, study allows comparison of the risk of dying in 1965 to risk of dying in 1985
Carcinogen—a substance that causes cancer. This term has been used confusingly in several ways—it may be a substance that causes tumors in particular species of animals. A human carcinogen is a substance that bears a cause and effect relationship to cancer in a person.
Cardiopulmonary Resuscitation (CPR)—the art of providing emergency assistance that supports the heartbeat and respiration in an individual under urgent and life-threatening circumstances, generally where heart activity and respiratory function has ceased.
Cardiovascular—heart and blood vessels, circulatory system
Cartridge-type Respirator—"gas mask" with charcoal filter containing cartridges that a worker wears to purify his air
Catastrophic Event—life-threatening situation
Causation—a factor leading to an event without which the event is unlikely to have happened
Caustic—an agent, particularly an alkali, that can destroy living tissue
Central Nervous System Depressant—blunts or depresses brain functions
Central Nervous System Dysfunction—effect on the brain and nervous system that impairs optimal function
Cephalosporins—class of antibiotics
Cerebrovascular Accidents (CVAs)—strokes
Chemical—a natural, manufactured or synthetic substance having a particular molecular structure and a Chemical Abstracts Service registry number (CAS); as the term is used in this book, a chemical may have harmful or beneficial properties and may be used as a drug or as an industrial compound; all living matter is composed of chemicals
Chemical Background—see Background, Chemical
Chemical Contact History (Incident)—a chemical contact history taken subsequent to an extraordinary incident such as a splash, spill or other accident
Chemical Contact History (Routine)—history of work with chemicals, specific details

Chemical Hepatitis—liver disorder caused by exposure to certain poisons or drugs

Chemophobia—fear of working with chemicals

Chemotherapy—treatment with chemicals (drugs) for a beneficial purpose

Chloramphenicol—an antibiotic not widely used today because of its relatively rare side effect of aplastic anemia

Chlorinated—contains chlorine

Chlorinated Pesticide—nickname for pesticides containing a chlorine group, not widely used in United States at present

Chlorinated Tap Water or Swimming Pool Water—water that has been disinfected with chlorine or chlorine-containing compound

Cholinesterase—an enzyme naturally present in humans, animals and insects which catalyzes the hydrolysis (breakdown) of choline esters such as acetylcholine. Acetyl choline transmits nerve impulses. In a person, cholinesterase inhibition can be evaluated from blood tests which measure the activity of the enzyme in red blood cells and in plasma.

Cholinesterase Inhibition—a substance which inhibits the activity of the enzyme, acetyl cholinesterase, see Pesticides, Cholinesterase Inhibiting.

Chorioretinitis—inflammation of the choroid and retina of the eye

Chronic—continuing or ongoing

Chronic Condition—a symptom or circumstance that persists for a period of time

Chronic Effect—a symptom or a finding which lasts a long time

Chronic Experiment—long term study

Chronic Experiment or Study—long-term studies in which animals are dosed for their lifetime or a significant portion of their lives

Chronic Exposure—exposures that continue over time

Chronic Obstructive Pulmonary Disease (COPD)—also known as Chronic Obstructive Lung Disease (COLD), disease process with specific diagnostic criteria including a history of persistent shortness of breath (dyspnea) on exertion and diminished maximum breathing capacity. Common conditions which result in COPD are chronic bronchitis and pulmonary emphysema, often from cigarette smoking.

Claims Adjusters—persons who evaluate and process insurance claims

Class—category of drug or chemical

COLD—Chronic Obstructive Lung Disease; see Chronic Obstructive Pulmonary Disease

Cold—common or laymen's term for an upper respiratory infection caused by a virus

Colic, Lead—see Lead Colic

Complaint—in a medical sense, used to describe a symptom or a concern; the reason for the visit to the physician.

Concentration—amount of something per unit volume. For a drug in the blood, it might be an amount of drug per milliliter of blood; the concentration of chemical is related to the dose and influences the response.

Contact—substances around us or encountered at work

Contaminants, Air—see Air Contaminants

COPD—see Chronic Obstructive Pulmonary Disease

Corticosteroids—a class of drugs which acts in a manner resembling that of the cortex of the adrenal gland, most corticosteroids in use today are synthetic compounds. When used for long periods of time they are associated with many serious side effects such as predisposition to infection, diabetes, demineralization of bone and gastrointestinal ulceration and hemorrhage. Corticosteroids may be used during severe asthmatic states which have not responded to other treatments as they are effective in abating these attacks; however, they should not be used more than one week or ten days in such patients as the propensity for asthmatics to become "steroid dependent" is high, and long-term use predisposes these patients to the life-threatening side effects.

CPS I and II—see Cancer Prevention Studies I and II

Cotinine—a metabolite of nicotine that is present in urine; useful in the determination of whether or not a person is presently smoking

Cubic Meters—35.3 cubic feet, a volume of air expressed m^3

Curing—as used in industry, the setting or hardening of a chemical reaction

Cutting Fluids—oil or water-based fluids used to cool and lubricate cutting tools during machining

CVA—see Cerebrovascular Accidents

Delayed Recovery Syndrome—failure to return to work after an injury, at a point in time when most people would have recovered

Dermatitis—skin irritation or inflammation

Dermatitis, Solvent—skin changes due to exposure to a solvent, typically consisting of drying, defatting and cracking of exposed skin

Desensitizing Antigens—as used by environmental illness philosophers, desensitizing antigens are generally not those used by conventional specialists in allergy. Patients receive items such as "diesel fuel," "petrochemicals" and "newsprint" to be taken sublingually (under the tongue) or by injection. These "treatments" are of no known benefit and their safety has not been proven.

Diabetes—a family of diseases of juvenile and adult onset where altered insulin results in abnormal glucose metabolism, many body organs and systems can be affected

Diagnosis—medical determination or opinion of an illness or injury

Differential Diagnosis—disease possibilities based on history and exam findings

Diffusion Capacity of the Lung—measures ability of air to pass across lung air sacs (alveolar membranes) and is a measure of integrity of functioning lung tissue

Disability—impairment or limitation

Disability Retirement—retirement for medical reasons not necessarily work-related

Distribution—where the substance goes in the body

Dose—amount of drug or chemical that enters the body

Dose-Response Concept (for drugs and chemicals)—the effect of a substance is related to the amount that is present. For a drug, an insignificant amount will have no effect, more will be beneficial and too much will be toxic. The "dose makes the poison."

Drug—chemicals intentionally taken into the body for a specific purpose or effect; chemical taken voluntarily; taking a drug is a voluntary chemical exposure

Drug Fever—an elevated body temperature or fever due to an allergic reaction to a drug

Drug Interaction—adverse situation arising from irrational drug combinations

Drug Side Effects—an unwanted drug effect; a toxic response to a drug. Drug side-effects are common but permanent effects are uncommon.

Drug-Seeking Behavior—falsifying or exaggerating symptoms for the prime reason of obtaining drugs

Duration—length of time

Dust—small particles

Dust, Organic—generally used to describe dust from vegetable or animal matter, as opposed to mineral

Dysesthesias—disagreeable or painful sensation such as pins and needles or crawling feeling of the skin

EBV—see Epstein-Barr Virus

Effect—an action of a substance in the body

Elephantiasis—inborn or acquired obstruction of lymph nodes leading to deformity of the affected part, most commonly lower extremities and scrotum

Elimination—how the body disposes or gets rid of a substance; drugs do not stay in the body forever.

Emergency Responders—persons such as emergency medical technicians (EMTs), firefighters, police, hazardous material workers (HazMats), or persons in a workplace or a community trained to respond to emergencies and render cardiopulmonary resuscitation (CPR)

Emission—escape into air such as a gas

Employer's Claim Coordinator—person in a company who handles injured worker claims

Encapsulating—fully enclosed

Endemic—native to the region

Endogenous—due to internal factors; for an analyte, a substance that is made within the body and is unrelated to a chemical exposure

Engineering Controls—preventing or correcting the escape of a substance or substances into the environment by mechanical design and devices

Entrained—enmeshed or incorporated

Environmental Allergy/Detoxification Center—unconventional sauna treatment for unknown toxins, not accepted as medical treatment for any known condition

Environmental Illness—as used in this book, a euphemism for an unconventional philosophy subscribed to by certain patients and practitioners wherein the cause of a person's woes is due to the "environment". Such philosophy is out of the mainstream of medicine and not generally accepted or confirmed by scientific evidence. The following buzzwords should alert the unwary to these claims: "environmental allergies, clinical ecology, immune system damage, permanent sensitivities (to the environment), psychic stress, brain damage, headaches, loss of intellect and immune disregulation."

Epidemiological—pertains to the science of epidemiology, studies, defines and explains factors that determine disease frequency and distribution

Epoxies—a synonym for epoxy or Epon resins; they are polymers containing more than one epoxide group, generally in a two-part system consisting of the uncured resin and curing agents also known as catalysts, accelerators, activators or hardeners. Depending on the product, there may be other components. Various components in the uncured systems or released during the curing process may produce irritation or sensitization of the respiratory system and/or skin (asthma, mucous membrane irritation and dermatitis).

Epstein-Barr Virus (EBV)—responsible for mononucleosis, and human tumors of the throat (Burkitt's Lymphoma)

Eruption, Skin—breaking out

Erythema—redness

Erythrocytes—red blood cells

Essential Hypertension—see Hypertension, Essential

Ethanol—alcohol; chemical structure CH_3CH_2OH

Exhausted—drawing emissions, gases, fumes and vapors away from the workers' breathing zone and work area by designing special ventilation that accomplishes this

Exposure—an amount of substance entering the body at a level over background, does not necessarily imply injury. It is important to determine whether or not an exposure occurred, since if there has been no exposure, there can be no injury.

Exposure, Acute—a sudden occurrence where an amount of substance over background encounters skin or enters some part of the body

Exposure, Chemical—see Exposure and Exposure, Toxic

Exposure, Chronic—over a period of time, low level amounts over background encounter skin or enter a part of the body. Note that use or contact with the substance does not necessarily imply exposure.

Exposure History—collection of historical facts to uncover evidence sufficient for a medical determination of whether an exposure occurred

Exposure, Route of—see Route of Exposure

Exposure, Toxic—an amount of a toxic substance over background which enters the body by inhalation or skin absorption, ingestion. Not a diagnosis unless verified by the Seven-Step Toxic Injury Verification Test (See Seven-Step Toxic Injury Verification Test).

Extermination—elimination of pests, often by the use of chemicals

Fat-Soluble—dissolves in fat

FDA—see United States Food and Drug Administration

FEP—see Free Erythrocyte Protoporphyrin

Fertilizer—an intentionally applied source of nutrients for plants, may be naturally occurring as manure from various animals, derived from other organic matter as by composting, or manufactured

Fetal Alcohol Syndrome—birth defects in some infants born to alcoholic mothers

Fiber—thread-like bodies or filaments

Fibrous—full of or formed of fibers; in the body, fibrous tissue may be scar tissue

Flu—nickname for influenza, often this term is used broadly to indicate a viral illness. Flu-like reactions, such as metal fume fever occur in workplace settings, but are not common.

Folliculitis—inflammation of hair follicles with pus formation, promoted by bacterial growth around hair follicles

Free Erythrocyte Protoporphyrin (FEP)—a blood test which measures a derivative of hemoglobin without an iron atom, increases in anemias and lead exposures

Fugitive Emissions—unintentional leaks into air from manufacturing processes, such as from improperly sealed pipe joints

Fumes—a term often used interchangeably with vapors, especially those having irritating or noxious qualities, smaller particles than dusts

Fumigation—the intentional use of chemicals (fumigants) into the air to eliminate pests

Gamma Glutamyl Transpeptidase (GGT)—a liver function test

Gastritis—irritation of the stomach

Genetic—inherited from one's parents

Germanium—a grayish white metallic element of the silicon group, used in electronics industry, of no known medical benefit

Gram—0.002 pounds; 0.03 ounce (avoir.)

Guillain-Barre Syndrome—this condition may occur after recovery from an infectious disease. It is characterized by polyneuritis with progressive muscular weakness of extremities that may lead to paralysis. If the acute period is uncomplicated, recovery is usually complete.

Hangover—don't ask; some or all of the symptoms of headache, nausea, big head, dizziness, sick feeling

Hay Fever—see Allergic Rhinitis

Health Effect—as used in this book, an unwanted response in the body; a definable or measurable impact on health

Heat Stress—a series of disorders caused by exposures to high levels of environmental heat, may result from too high a temperature, excess fluid and salt loss, inability to sweat. Promoted by too much of a workload at too high a temperature, especially by encapsulating and impermeable clothing.

Heat Stroke—one of the severe forms of heat stress disorders consisting of abnormally rapid rise of body temperatures to excessive levels followed by collapse, delirium, struggling and convulsions. May be fatal, requires immediate careful treatment.

HEENT—head, eyes, ears, nose, throat

Hematoma—blood clot

Hematuria, Microscopic—trace blood in urine visible under a microscope but not to a person

Hemlock—a family of poisonous plants, small amount of which when ingested can be fatal. Hemlock is believed to have been the poison administered to Socrates.

Hepatitis—inflammation of the liver due to viral infection, drug reaction or toxin; usually manifested by jaundice, liver enlargement; fever and other systemic disorders are usually present

Hepatitis, Fulminant—marked by sudden onset of nausea and vomiting, chills, high fever, severe and early jaundice, convulsions, shock, deep coma, and death usually within 10 days

Herniated Disc—rupture or herniation of intervertebral disc

HERP—a computation factor devised by Ames, McGaw and Gold, Science 1987 which they define as human exposure dose/rodent potency dose

Highly Reactive Compounds—chemically unstable chemicals, fast-reacting to become something else

Histamine Blockers—a nickname for histamine receptor antagonists. These drugs are used to treat ulcers and other gastrointestinal conditions but have many potentially serious side effects and a propensity for drug interactions.

History of Present Illness (HPI)—medical background relevant to the reason that the person presented for examination

Hot Spots in a Manufacturing Facility—these are locations or work areas plagued by spills, frequent upsets, fugitive emissions, maintenance problems, frequent breakdowns, worker complaints, worker problems and waste or waste generation

Hyperkeratosis—overgrowth of the horny layer of the skin (epidermis)

Hypertension, Essential—high blood pressure, genetic predisposition, of unknown cause

Hysteria, Mass—see Mass Hysteria

Hysterical—excited or irrational emotional reaction

IARC—see International Agency for Research on Cancer

Id—"me, too", one body part reacts to disease in another

Illicit—illegal; pertains to "street" drugs or drugs used for "recreational" purposes

Immune System—complicated body processes that protect against infection and tumors but sometimes overreact and cause allergies or illness

Immunosuppressive—acting to suppress the body's natural immune response

Impervious Gloves—gloves which will not allow anything to penetrate

Impotence—inability of a man to achieve or maintain an erection

Indoor Environments—enclosed or partially enclosed structure; a room or building

Industrial Hygiene History—history of routine and extraordinary measurements concerning a particular job, process or work area

Industrial Hygiene Surveillance—concentrates on workplace airborne threshold limit values, conducts air sampling, skin wipes and wipes of the work surfaces. These data can be used to estimate exposures.

Industrial Hygienists—specialists in measurement of air contaminants, obtain information from which chemical exposures can be calculated, recommend safety equipment to prevent exposure, usually certified in this field (CIH)

Inert Ingredient—an ingredient that may not have activity for the specific use of a product like a solvent in a pesticide, but may *not* be inert in people

Infectious Agents—bacteria, fungi, viruses, etc.

Inflammatory Cells—the body's security force, normally eliminates foreign substances

Ingest—take by mouth, such as swallowing a pill or by eating and drinking, for example, at a dust-contaminated work bench

Ingestion—see Ingest

Inhalation—see Inhaled

Inhaled—taking into the body by the respiratory route, breathing in

Injury—measurable unwanted effect

Insoluble—undissolvable

Inspected—looked at, observed

International Agency for Research on Cancer (IARC)—an organization of the World Health Organization (WHO) that organizes experts into working groups for the evaluation of carcinogenic risks to humans

Intoxicating—having a toxic effect

Intoxication—unwanted health effect or feature of poisoning

Intoxication, Alcohol—see Alcohol Intoxication, Clinical; see Alcohol Intoxication, Legal

Irritants, Chronic—those substances that can stir up chronic inflammation

Irritation, eye—redness, tearing or burning of eyes

Isocyanates—family of chemicals used in the application of polyurethane coatings such as auto body spray paints or the manufacture of polyurethane foam products. RAST tests may be positive in persons who have developed isocyanate-induced asthma.

Isopropyl Alcohol—rubbing alcohol, a solvent

Itis—suffix, denotes inflammation

Laryngospasm—spasm of the muscles of the larynx associated with a sensation of choking

LD_{50}—the amount of chemical that kills half the animals

LD_{LO}—the lowest dose that kills an animal

Leach—dissolves out from

Lead—a metal; an element with no essential use in the body, present in the body as a contaminant

Lead Burden—total amount of lead in the body estimated from indirect tests

Lead Colic—severe abdominal cramps from lead poisoning

Lethal—causes death

Light-Duty Jobs—jobs that allow employees to begin work again within medical restrictions

Liter—about a quart

Liver Function Tests—blood tests that measure substances that occur naturally in the body which reflect various functions of the liver. These may include blood proteins such as albumin and globulin; alanine aminotransferase (SGPT), aspartate aminotransferase (SGOT), bilirubin, prothrombin time, alkaline phosphatase and gamma glutamyl transpeptidase (GGT). While liver dysfunction commonly occurs as a result of obesity, alcohol, drug use and the like, alteration of liver function tests as a result of workplace exposure in a person is unusual. Liver dysfunction is commonly described on MSDS as a result of findings in animal experiments under circumstances which generally do not apply to workers.

Long-term Medication—a course of treatment that commits a person to receiving medication for more than a week or two

Lost Time From Work Injury—sometimes defined by law, consisting of an absence from work for more than one to three days

Lower Respiratory Passages—small bronchi and bronchioles

Lupus—denotes several forms of an autoimmune disease of connective tissue which may affect the skin, joints, kidneys, nervous system and mucous membranes. A characteristic butterfly rash or erythema may be present across the bridge of the nose and on the cheeks.

Lymph Nodes—a rounded accumulation of lymphatic tissue located in superficial and deep areas of the body; the cells in the lymph nodes perform immune functions

Malaise—weakness, a feeling of being unwell

Malfunction—upset or breakdown

Mass Hysteria—groups of people who overreact without valid reason

Material Safety Data Sheets—specific chemical information sheets that are required to be provided by the manufacturer of a chemical product. These sheets provide information about physical properties, ingredients, fire and explosion data, health hazards data, effects of overexposure, reactivity, procedures for handling spills and leaks as well as information about special protection and precautions. Examples of MSDS and a further discussion can be found in Chapter 19.

Medical History—discussions with the patient to obtain specific information and facts concerning present and past health

Medical Monitoring—see Medical Surveillance

Medical Surveillance—medical examinations at specified intervals including laboratory evaluations designed to protect workers and prevent work-related injuries; a program that prevents health problems by constant monitoring. When specific tests are not available, medical personnel detect transient symptoms before permanent injury occurs; documents effort and health status. When specific tests are available, such as for lead, the amount in the blood is measured at intervals and the person is removed from exposure *before* harmful levels are attained.

Medical Surveillance Program for Toxics—identifies potentially hazardous jobs; conducts pre-hire medical and competency screening; obtains samples for baseline biologic monitoring; records chemical contact history; records exposure history; special responses to exposures at levels sufficient to produce an injury; overreacts to questions, incidents and injuries; interval surveillance; industrial hygiene monitoring; worker education and participation in medical surveillance; termination or retirement evaluations

Medication—drugs used to treat a condition or taken for a beneficial purpose

Medication Side Effect—an unwanted drug effect; a toxic response to a drug. Drug side-effects are common but permanent effects are uncommon.

Memory—ability to recall past knowledge

Meniere's Disease or Syndrome—a recurrent and often progressive group of symptoms which include progressive deafness, ringing in the ears, dizziness and the sensation or fullness or pressure in the ears

Mental Status—usually consists of orientation to time, place, person and state of mind, and affect (emotions)

Metabolism—alteration in the body of one substance into another, called a metabolite

Metabolite—the product of a substance resulting from alteration in the body; the result of conversion of one substance to another by the body

Metabolize—convert one substance in the body to another

Microgram (μg)—one thousandth of a milligram or one millionth of a gram

Micrograms per Cubic Meter—one millionth of a gram per cubic meter of air (μg/m^3)

Micrograms per Deciliter (μg/dL)—(one-tenth of a liter, about 1/10 quart); this is a concentration of a substance, such as lead in blood

Microscopic Hematuria—red blood cells in the urine only evident when examining the urine under a microscope, the patient is usually not aware of this finding. It may be benign or of no health consequence especially in young men, or it may be indicative of a more serious medical problem. In some patients, it may indicate conditions such as a kidney stone or a tumor. Irrespective of the cause, it is not usually work-related.

Migraine Headache—paroxysmal attacks of headaches, frequently one-sided, usually accompanied by distorted vision and gastrointestinal symptoms. Migraine headaches are thought to be the result of dilation of arteries of the head outside the brain (extracerebral cranial arteries).

Milligram (mg)—one thousandth of a gram

Milliliter (ml)—one thousandth of a liter

Mill Tailings—rock dust remaining after an ore of interest, such as lead, has been removed by crushing (milling) and concentration

Miscarriage—spontaneous abortion, loss of a fetus

Misdiagnosis—incorrect medical determination; incorrect interpretation of a patient's symptoms or disorder

Mucus—a sticky fluid secreted by mucous membranes and some glands

Mucous Membranes—membranes lining passages and cavities communicating with air such as the nose, mouth, respiratory tract and eyes

Mucous Membrane Irritation—consists of some or all of the following signs and symptoms with respect to the mucous membranes: watering and redness of the eyes, burning and running of the nose, burning and scratchiness of the throat and cough. These symptoms are usually transient and abate after a person is removed from exposure.

Narcotics, Opiate—commonly found in pain, headache and cough suppressant medication with effects similar to opium. There is a high propensity for addiction and abuse especially in patients with musculoskeletal and orthopedic problems. Narcotics should be used only when absolutely necessary and for short periods of time. These drugs should not be prescribed for patients with chronic pain as there is a high propensity for abuse.

Nasal Polyp—see Polyp, Nasal

National Institute of Occupational Safety and Health (NIOSH)—the research arm of OSHA, makes recommendations to OSHA

Naturopath—non-medical practitioner or nutritionist, licensed in some states

Negative Experiments—nothing bad happened

Neurofibromatosis—condition in which there are tumors of various sizes on peripheral nerves, also see Von Recklinghausen's Disease

NIOSH—see National Institute of Occupational Safety and Health

Nonsteroidal Anti-inflammatory Drugs—class of drugs often prescribed for arthritic or muscular complaints

Noxious—unpleasant, unwholesome or harmful

Occupational Medicine Surveillance History—specific facts of past exams concerning worker health and exposures

Occupational Safety and Health Administration (OSHA)—a U.S. regulatory agency; provides regulation for protection of worker health to include items such as safe air levels, medical programs, blood tests, safe work habits and personal protection. Individual states may have regulations that are more stringent.

Odor—perception of a smell, often with great variability among individuals

Off-Gassed—gaseous fumes emanating into the air, often during curing or baking

Organ—body parts

Organic Dusts—dusts containing or derived from organic matter such as bean dust or wood dust

Organic Volatiles—nickname for substances that readily evaporate into the air, such as many solvents

Organically Grown—produce raised without the addition of manufactured chemical substances

Organophosphate Cholinesterase-Inhibitor—see Pesticide, Cholinesterase Inhibiting

Orientation—whether a person is aware of where he/she is, their name, date, time and place

OSHA—see Occupational Safety and Health Administration

Ounce (oz.)—(avoir.) 28.35 grams, 1/16 pound

Overdose—an excessive amount of drug or chemical entered the body

Overexposure—an excessive amount of substance(s) entering the body approaching levels that *may* cause harm; a medical determination is needed in such cases to establish whether or not a harmful effect or injury has occurred

Palpated—felt

Papular—raised

Parkinson's Disease—this chronic nervous disease is characterized by a fine slowly spreading (pill rolling) tremor, muscular weakness and rigidity and a characteristic peculiar shuffling gait. The face becomes expressionless and speech slow. Recovery is rare but condition may improve with drug treatment.

Past Medical History (PMH)—history of prior illnesses

Penicillin—antibiotic

Perchloroethylene (PCE)—chlorine-containing solvent used in drycleaning and industry

Percussed—tapped

Peripheral Neuropathy—altered function in nerves in arms and/or legs. Peripheral neuropathy may have many causes. It may be a symptom presented by a lead-intoxicated person, and can be evaluated both by physical examination and by testing peripheral nerves.

Permanent Disabilities—an injury that would be compensable and rated as such

Permanent Injury—a lasting effect

Personal Protection—device(s) a worker would wear on his person, to prevent an exposure

Perturbations—disturbance

Pesticide—any substance used to control nonhuman pests; herbicide, fungicide, insecticide, miticide or other item designed to kill unwanted weeds, fungi, insects, mites or other pests on crops, people, pets and/or indoor and outdoor environments

Pesticide, Cholinesterase Inhibiting—a pesticide which inhibits the enzyme cholinesterase. Structurally, these compounds tend to be of the organophosphate or carbamate class. Organophosphate intoxication can be assessed by measurement of the activity of the cholinesterase enzyme in a person's red blood cells and plasma.

Pesticide Intoxication—an amount of pesticide that has entered the body and caused symptoms

Pesticide Poisoning—see Pesticide Intoxication

Pharmacokinetics—the pathway of a drug in the body from entry to exit, includes absorption, distribution, metabolism, effect and elimination

Pharmacology—scientific study of drugs

Phlegm—thick mucus

Physician-Toxicologist—a medical doctor licensed in medicine and experienced in human toxic reactions

Pneumonia—inflammation of the air sacs of the lung generally due to infection with bacteria, viruses or other infectious agents. While a chemical pneumonitis is possible following significant exposures to severe irritants or highly reactive chemicals such as methyl isocyanate, chemical pneumonitis is uncommon and generally does not occur without an accident such as a spill or a release.

Pneumonitis—inflammation of the lungs, sometimes used interchangeably with pneumonia

Poison—any substance which when absorbed by a living organism destroys life or injures health. This term is often applied to substances which act rapidly when taken in a small quantity.

Polychlorinated Biphenyls (PCBs)—synthetic flame and heat retardant of complicated chemical structure containing multiple (poly) chlorine atoms. The use of these compounds in the U.S. has been phased out because of their stability in the environment.

Polyp, Nasal—a stalk-like tumor of the mucous membrane of the nose

Post-Infectious—begins following an infection (e.g., sinusitis, bronchitis, pneumonia)

Potent—powerful; in animal studies of toxicity, a chemical that kills the most animals at the lowest dose is the most potent

Potentially Hazardous Jobs—job where the worker may encounter chemical exposure or highly reactive, explosive chemicals, or chemicals readily absorbed through the skin; or because there is potential exposure to large quantities of a low hazard substance

Pound (lb.)—(avoir.) 16 ounces; 453.6 grams

ppb—parts per billion

ppm—parts per million

Pre-employment Physical Examination—a baseline medical examination generally including laboratory evaluation at the start of a new job

Pre-hire Exam—see Pre-employment Physical Examination

Present Complaint—the reason the person seeks medical advice

Present Medication—the medication the patient is currently taking

Presents—person manifests symptoms (e.g., runny nose)

Primary Provider—physician or clinic selected by the employer to provide medical care to their injured employees

Process—steps it takes to manufacture products; chemical and mechanical procedures

Process Intermediates/Streams—chemicals that are past raw materials, which have been reacted but are not yet finished products

Process Specifications—the "recipe" and desired result of a chemical process

Prospective Survey or Study—a group of people are identified and then observed to see what happens to them

Protective Devices—equipment used in the worker's area or on his person designed to minimize the chance of a chemical exposure, may include factors such as ventilation, segregation of the worker from the process and protective gear for eyes, face, body, hands, feet and respiratory systems

Protective Measures—see Protective Devices

Pulmonary Edema—fluid-filled lungs, may result from varied causes including heart failure, fluid overload or from a toxic lung injury such as from hydrogen sulfide gas poisoning

Pulmonary Function Tests—tests that measure various lung functions including airway functions, lung volumes and gas exchange

Putative Agents—suspected toxic mischiefmaker

Radiation—radioactive materials, includes substances such as in sunlight; radiation used in diagnosis and treatment, and nuclear radiation

Radioallergosorbent Tests (RAST)—allergy tests from blood which measure IgE antibodies to specific foods (e.g., peanut), pollens, weeds, etc.

RAST—see Radioallergosorbent Tests

Reactive Chemical or Material—chemical or chemical mixture which readily interacts with other chemicals, people or the environment

Rebuttable Presumption—a regulation, generally for public servants, that presumes that certain disorders such as heart conditions are job-related unless rebutted (disproved) by evidence to the contrary

Recreational Drug Use—non-medical drug use, street or prescription drugs

Red Blood Cells—erythrocytes

Relative Cancer Risk—assumes the risk of cancer without an exposure is 1.0, then compares the risk of an exposure to a carcinogen. If there is ten times more chance of developing cancer in the exposed group, the relative risk would be 10.

Relative Risk—the incident rate of a disease in the exposed group divided by the incident rate of disease in a non-exposed group. For example, if a study is designed to determine the incidence of lung cancer between smokers and non-smokers, the incidence of lung cancer between smokers would be divided by the incidence rate of lung cancer in non-smokers, see Relative Cancer Risk.

Release—unplanned or undesirable emission of chemicals into the environment

Reproductive Study—a particular type of chronic study which measures effects of drugs and chemicals on various aspects of animal reproduction and on their offspring

Residual—remaining

Respirator, Cartridge-type—"gas mask" with charcoal filter-containing cartridges that a worker wears to purify his air

Respirator, Fit-test—a test, often involving banana oil, to ensure that a respirator is fitting properly and not allowing contaminants to be breathed

Respiratory—concerned with getting air to the lungs; pertains to organs concerned with breathing

Respiratory System—organs involved in the interchange of air consisting of the nose, pharynx, larynx, trachea, bronchi and lungs

Review of Systems (ROS)—review of the major organ systems of the body, such as nervous system, respiratory system, circulatory system, etc.

Risk Managers—persons who specialize in preventing untoward events, promoting safety and minimizing risk, usually employed by insurers or large companies

Robotics—use of robots in manufacturing

Route of Exposure—portal of entry of a substance into the body, as by mouth, inhalation, injection, skin absorption or some combination of such factors

Safety Professional—person in a company who is responsible for toxics management and worker safety

Sarcoid—a disease of unknown cause characterized by widespread granulomatous lesions that may affect any organ or tissue of the body. Liver, skin, lungs, lymph nodes, spleen, eyes and small bones of the hands and feet may be affected.

Sebaceous Glands—oil-secreting glands of the skin; usually associated with hair follicles

Secondary Infection—infection caused by something different than the one causing the primary infection

Seizure Disorder—convulsive disorder

Sensitive—made susceptible to a specific substance

Seven-Step Toxic Injury Verification Test—This test is conducted by asking the following seven questions: 1. What is the person's motivation for the visit? 2. What is the diagnosis? 3. Did the person sustain chemical exposure? 4. Was the exposure of sufficient intensity to produce an injury? 5. Is there evidence that an injury was produced? 6. Is the injury likely to be a result of the exposure? 7. Does the conclusion make sense? In order to pass the test, the following answers are required: 1. The motivation for the visit was sound. 2. The symptoms or diagnosis are consistent with an exposure. 3. The person sustained the chemical exposure. 4. The exposure was of sufficient intensity to produce an injury. 5. An injury resulted. 6. The type of injury logically results from the exposure. 7. The conclusion makes sense.

Shingles—more formally known as Herpes Zoster, it is an acute infectious disease caused by the varicella-zoster virus which is responsible for chickenpox. This condition is characterized by inflammation of the posterior root ganglia of a few segments of the spinal or cranial nerves leading to a painful, vesicular eruption along the course of the nerve. Usually the blisters only occur on one side.

Side-effects—an unwanted drug effect; a toxic response to a drug. Drug side-effects are common but permanent effects are uncommon.

Significant Aggravation—a situation where workplace factors worsen or prolong symptoms of a pre-existing non-industrial condition, such as asthma

Sinusitis—inflammation of the linings of the sinuses often due to or associated with bacterial infections, allergies, or chronic irritants as with the use of tobacco or street drugs. May also be associated with workplace irritants

Skin Contact—a substance has made direct contact with the skin, such as by a splash or direct application with fingers. Excessive amounts of vapor in air may also contact the skin.

Skin Wipe—wiping the skin with a cloth and extracting the chemicals from it to determine what was present on the skin. May be of use in assessing exposure in the event of a spill, splash or release.

Smoke—air contamination resulting from burning

Soluble—dissolvable, dissolves

Solvents—substances that dissolve chemicals, oils, fatty material or grease

Soot—visible particles resulting from combustion or burning of coal, wood, oils or other substances used as fuel

Species—forms or kind, a breed or family of animals; a type of chemical

Status asthmaticus—severe sustained asthma resistant to treatment

Substance Abuse—alcohol, tobacco, and drug overuse and addiction

Substances of Abuse—substances such as street drugs (marijuana, cocaine, heroin, PCP, amphetamines, etc.), tobacco or alcohol, that would not ordinarily be prescribed by a physician; may also imply prescription drug or street drug overuse and addiction in some cases; substance abusers commonly abuse prescription drugs as well as over the counter medication including medication for pain, tranquilizers, antianxiety drugs, headache medication, cough suppressants, caffeine, diet preparations containing pseudoephedrine and phenylpropanolamine.

Symptom—a perceptible change in the body or its function; what a person complains of

Symptom, Acute—a perceptible change in the body or its function that comes on suddenly

Symptom, Chronic—a perceptible change in the body or its function that lasts for a period of time

Symptom, Mucous Membrane Irritation—red watery eyes, runny nose, scratchy throat, possible cough, for example, from a heavily chlorinated swimming pool. Symptoms last until the person leaves the pool and then disappear.

Symptom, Skin Irritation— redness, smarting and burning are common complaints

Symptom, Solvent Overexposure—symptoms may range from feeling high, light-headed, dizzy, nauseous, headache or drunkenness. Generally will not occur unless workers are in an enclosed space in which a high concentration of solvent is present without appropriate protective equipment. Remember the effects of alcohol. In non-catastrophic cases, when removed from exposure, the effects disappear.

Symptom, Transient—temporary symptom

(TDI)—see Isocyanates

Tardive Dyskinesia—this condition, characterized by abnormal and uncontrollable (choreiform) movements of the mouth and face, sometimes of the extremities, may be caused by high doses of phenothiazine medication for psychiatric patients. Older persons, especially women and those with brain injury, are particularly susceptible. Unfortunately, the problem does not disappear when the drug is discontinued and resists standard treatment for movement disorders.

Teratogenic Study—measures fetal malformations and birth defects

Testicular Atrophy—a wasting and decrease in the size of the testicles

Therapeutic Drug Levels—tests which measure amount of drugs, generally prescribed by a physician, in blood and urine samples

"Tight Buildings"—generally, buildings with centralized heating, ventilation, and air conditioning systems and sealed to be energy efficient which do not have adequate ventilation for the number of occupants and purpose of use. The problem is usually corrected by improving the intake of fresh air.

Time Weighted Average Threshold Limit Values (TWA TLV)—designated safe air levels of particular substances based on averages over a 40-hour work week for a working lifetime

Toluene Diisocyanate (TDI)—see Isocyanate

Tinea pedis—athlete's foot; fungus infection of the feet

Total Exposure Assessment Monitoring (TEAM)—studies conducted by US EPA which identified and measured chemical background in air, water and breath

Toxic Effect—a measurable response after an exposure, not necessarily a compensable injury

Toxic Injury—criteria includes: sound motivation for the visit; symptoms or diagnosis consistent with exposure; exposure occurred; during exposure, enough was absorbed to cause injury; injury resulted; the type of injury logically results from exposure; scenario makes sense; also see Seven-Step Toxic Injury Verification Test

Toxic Injury Verification Test—devised in this book, an original concept, see Seven-Step Toxic Injury Verification Test

Toxic Injury, Workers' Compensation—results in a disability that can be defined and is compensable

Toxic Levels—the concentration of a chemical above which adverse health effects can occur

Toxic Response—toxic effect, reaction in the body to a toxic substance, not necessarily an injury

Toxicokinetics—the action of a chemical or unwanted properties of a drug in the body from entry to exit, includes absorption, distribution, metabolism, effect and elimination

Toxicologist—a person with an interest in toxicology, not necessarily a medical doctor, toxicologists may have training from experience, master degrees, or Ph.D. degrees; persons who study toxic reactions in animals or people

Toxicology, Animal—the study of the effects of substances in animal experiments; generally in rats and mice, substances are administered in large overdoses; the results of animal experiments cannot readily be applied to humans

Toxicology, Human—the study of substances that are harmful to people

Toxics—substances with potentially harmful properties; nickname for toxicology

Toxin—a toxic or poisonous substance

Trace quantities—extremely small amounts of a substance not associated with any biologic reaction or health effect. Trace quantities are too low to cause any effect on health but their presence can be detected by sophisticated technology

Tricyclic Antidepressant—a drug class containing three rings that is used to treat depression

Tumor—an unwanted growth not necessarily malignant or cancerous

TWA TLVs—see Time Weighted Average Threshold Limit Values

Unconventional Philosophies—"environmental illness"; "environmental allergy," "clinical ecology," immune system damage, permanent sensitivities; psychic stress; brain damage, headaches, "loss of intellect"; immune disregulation

United States Environmental Protection Agency (US EPA)—regulatory agency for indoor and outdoor air pollution, pesticide use, waste disposal, toxic substances and sets federal drinking water standards; states may have more stringent regulations

United States Food and Drug Administration (US FDA)—regulatory agency that assures the claimed benefits of drugs, and protects the public from undue risk of toxic effects

Untoward Event—adverse or undesired occurrence such as a spill, splash or release; an accident

Upper Bound Background—highest amount still considered "normal"

Upper Respiratory Passages—nose, throat, trachea, large bronchial tubes

URI—upper respiratory infection

Urinalysis—examination of the urine. The urine is examined for specific gravity, and the presence of protein, glucose, blood cells and evidence of infection or tumor. This is a routine urinalysis. The urine can also be analyzed on special studies for evidence of various chemical exposures.

Vapors—gaseous state of a substance

Vesicular—blister-like

Viral—pertaining to a virus

Vital Functions—respiration, circulation, oxygenation, blood pressure, body temperature and other basic functions which ensure the integrity and health of a person

Vital Signs—temperature, blood pressure, pulse, respirations

Index